CHAMELEONS
and Other Quick-change Artists

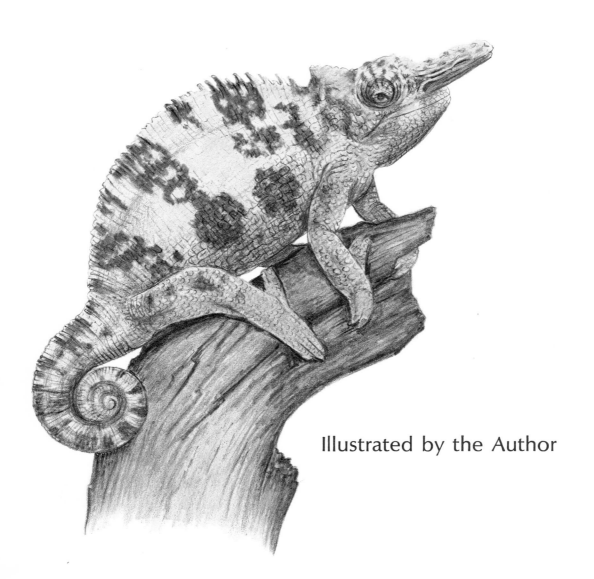

Illustrated by the Author

CHAMELEONS
and Other
Quick-change Artists

HILDA SIMON

DODD, MEAD & COMPANY · NEW YORK

ISBN: 0-396-06801-4
Library of Congress Catalog Card Number: 73-1656
Printed in the United States of America

Contents

Illustrations

ILLUSTRATIONS

Preface

THE CRAB that turns purple with rage; the lizard that announces its victory over a rival by changing to a vivid red; the octopus that blushes a shocking pink as it tries to hide after being disturbed—who has ever heard of them? Very few people, I would imagine, with the exception of the naturalists, both amateur and professional, whose business it is to know such things.

That does not mean, of course, that the capability of certain animals for rapid color change has remained unknown and unnoticed; on the contrary, some of these animals have become quite famous because of their quick-change artistry. Until quite recently, however, modern scientific studies of such color changes have centered predominantly on the biochemical, physiological aspects of the phenomenon. The part played by outside factors such as light and temperature, the reactions that result from artificial stimuli—injections of chemicals, electrical shock—were the areas of research that prevailed almost exclusively until a few decades ago. The philosophy of utilitarianism in biology that spread in the wake of Darwin's evolutionary theory quite naturally concentrated upon physiology and biochemistry rather than upon psychology and behavior. Experiments involving live animals, especially those of the "lower" classes, almost always ended in the subjects' death. Thus animals were blinded, decapitated, and electrocuted. Their spinal chords were severed, their limbs amputated, their brains removed, and nerves in various parts of their bodies were cut in efforts to gain the desired insights. At the same time, very little attention was paid to their moods and emotions and behavior, mostly for fear of being accused of anthropomorphism.

Over the past few decades, however, animal behavior studies have been accepted as an important part of biology. Such prominent naturalists as Dr. Karl von Frisch, Dr. Adolf Portmann, and Konrad Lorenz helped pioneer a new and much broader concept that included, among other things, the field of ethology.

Popular interest in animal behavior, slow to take hold at first, was accelerated by the realization that man and his technology have pushed the delicately interbalanced ecological community of our earth to the brink of disaster. The fact that many species of animals have been exterminated through the effects of technological progress, and that scores of others are today in danger of becoming extinct, has caused increasing numbers of people to concern themselves with the plight of the earth's wildlife. With that concern came the interest in the ways, the habits, and the behavior of these animals, and the discovery that a rich harvest of material on those subjects was available, dealing with everything from insects to whales. There was the fantastic "dance language" of the bees as revealed by von Frisch; there was Lorenz conversing with wild geese; there was Jacques-Yves Cousteau recording the "song" of the whales. We have seen lions that were born free voluntarily choose human beings as their companions and friends; have witnessed naturalists living near, and moving about with, entire families of the "ferocious" gorilla in the African jungles, and following tigers about—armed only with a camera—in the jungles of India. Along the way, old and stubborn assumptions, such as the alleged invariable viciousness of the killer whale, or that of the much-maligned wolf, were demolished forever. The entire subject of animal communication is yielding new and often astonishing results, and scientists even now are intent upon breaking the code of the porpoises' "language" so that men may actually talk to these animals.

The rapid color changes that are the subject of this book also represent a kind of language employed by certain animals to express and communicate moods or conditions—such as anger, fear, or sexual excitement—not through the more conventional means of sound or even movement, but through a variety of colors and color patterns. Continued patient observation is piecing together an interpretative code that helps us understand this color language in an increasing number of species.

When I embarked upon the study of this subject several years ago, I began to keep as pets a number of small reptiles and amphibians of species among which rapid color changes are pronounced. One of these animals was a grass frog named Maximilian, Max for short, who has become as one of the family over the course of the past two years. Observations of Max as well as other individuals of his kind have yielded valuable insights into the patterns of color change in this species. Almost more important, however, was the fact that this frog displayed a behavior that made him a distinct personality within his species. It is touching and challenging at the same time to find that such small and primitive creatures are capable of feelings, of a heightened zest for living, of individual differences that show them to be more than just "specimens."

Because of the wealth of information he supplied, and because of the pleasure he continues to give me, I therefore dedicate this book to Max and all his fellow creatures, small and large, whose continued survival in their particular environments depends upon mankind and its willingness to protect and preserve enough of this environment to give all the animals of this earth a chance to continue to share it with us.

CHAMELEONS
and Other Quick-change Artists

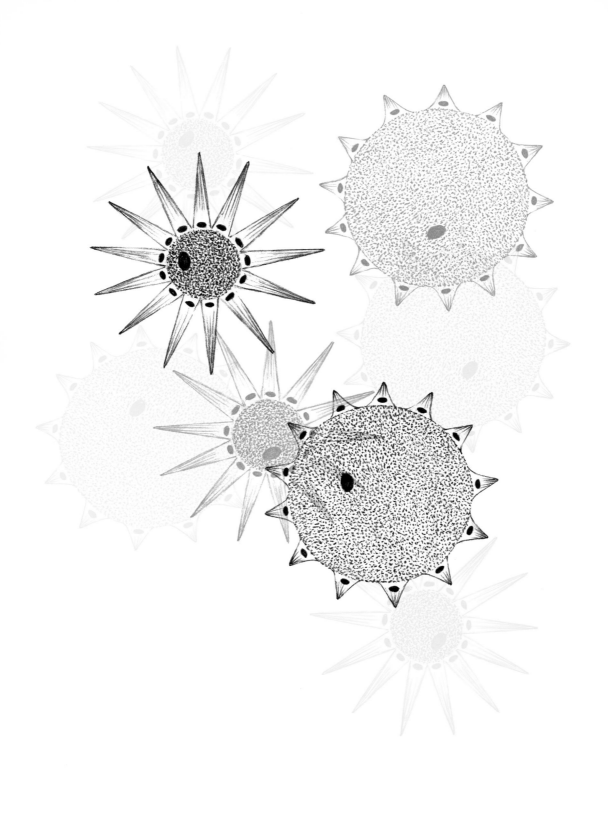

1. Changing the Colors

IF THERE is a single fact about any group of reptiles that almost everybody finds attractive, it is that chameleons change color. Of all the much-maligned reptiles, snakes are disliked and feared most, while lizards are often viewed with a suspicion that they are merely snakes on legs. Crocodiles and alligators admittedly have personalities that would not tend to endear them to anyone. This leaves turtles as the only reptilians that most people consider reasonably inoffensive and acceptable—with the exception, of course, of the famous chameleon, which in any case is hardly ever thought of as a reptile. Nor would the average person regard chameleons as lizards. The only thing that immediately comes to mind in connection with chameleons is their ability to change color—a trait frequently associated with the belief that these animals can make themselves invisible at a moment's notice. That notion, in turn, is based upon the supposition that the chameleon changes its color only to conceal itself and that an almost unlimited range of colors is available to this little reptile, permitting it to "melt" into backgrounds of every conceivable shade and hue. These assumptions, which were long ago shown to be partly inaccurate and partly downright erroneous, are still widely believed today.

Because of its legendary color-change artistry, the chameleon has been a source of fascination for thousands of years. To what extent this

17

lizard has captured the imagination is apparent from the fact that the word "chameleon" has found its way into many languages as a descriptive term. In English, German, French, and Russian—to name only a few—it is used as a synonym for a fickle and unreliable person, one who can "change his colors" instantaneously, adopting principles and opinions that are in sharp contrast to those held earlier.

An animal capable of rapid color changes was undoubtedly of special interest to the ancients because none of the domesticated animals and pets traditionally associated with man throughout his history possessed this curious ability. In mammals and birds, as well as in practically all mature insects, the colors and patterns that distinguish the various species are located in such special surface tissues as hairs, scales, and feathers, which by their very nature do not lend themselves to quick, chameleon-type color changes. Although a few species among

An ermine in its brown summer fur, and in its pure white winter pelt.

Ptarmigans change from a mottled brown feather dress to an all-white one in the fall.

these groups that may be considered, to a limited degree, exceptions to this rule, it holds true for the overwhelming majority. Normally, for instance, any fur-bearing animal can change its color only by the lengthy process of replacing hairs through new growth of a different color. Even this type of color change is relatively rare, found predominantly in northern regions among animals, with differing summer and winter pelts. Outstanding examples are the arctic fox, the variable hare, and several northern species of weasel known as ermine, whose pure white winter fur with its black-tipped tail is highly prized. All these animals gradually replace their brown or grayish summer fur with white hairs over a period of several weeks in the fall, and then repeat this process in reverse in the spring.

On a much wider scale, a similar process occurs in birds. A great many birds have distinctly differing summer and winter plumages. The

The immature little blue heron's white plumage is replaced by dark feathers in the adult bird.

arctic grouse replaces a feather dress of mottled brown and gray with a pure white plumage; the seasonal garbs of many other birds do not show that much of a contrast. However, the young of a great number of species have a juvenile plumage that is quite unlike the colors and patterns of the adult bird, and the males of many species grow a special feather dress for the mating period only. In all cases, the complete covering of contour, or outer, feathers—hundreds or even thousands of them—must be replaced, for these fully formed, "dead" tissues are incapable of any changes except those caused by wear and tear and fading, and all the pigments that produce the bird's distinctive colors and patterns are found in the contour feathers only.

At a superficial glance, those birds endowed with iridescent colors might be considered an exception, for their colors do change. At least they appear to change to the observer, who may be dazzled by the shifting hues of great purity and brilliance found in groups such as hummingbirds, sunbirds, glossy starlings, and many others. But iridescence is a very special and strictly optical phenomenon. Based upon interference, a process of selective reflection and absorption of certain wave lengths by submicroscopic tissue structures in birds and in numerous other animals, these colors shift—and may even disappear tem-

The male scarlet tanager has to undergo a mottled phase in the fall as he exchanges his brilliant breeding plumage for the dull colors of winter.

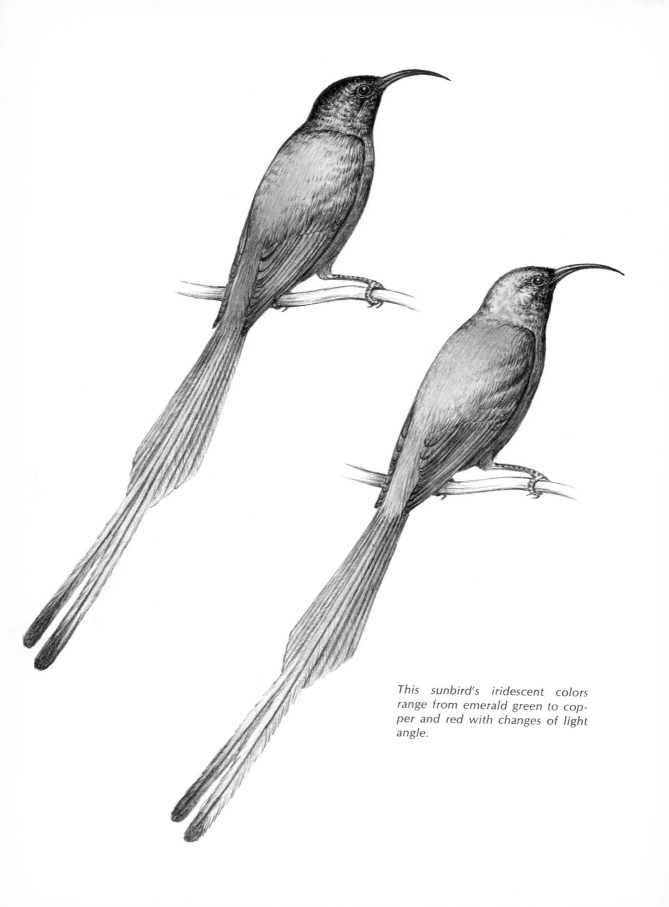

This sunbird's iridescent colors range from emerald green to copper and red with changes of light angle.

porarily—with any change of light angle or of the observer's position. Because of their structural, nonchemical origin, iridescent colors are remarkably stable and not subject to fading, usually remaining brilliant almost indefinitely even after the animal's death. Thus an observer looking at a mounted specimen in a museum will be able to see the same pure glittering hues, shifting with every change of position, as he would in a living animal.

In the huge class of insects, the rules governing the colors of the mature individuals are generally even more inflexible than among birds and mammals. Once the larval growth is completed, the colors and patterns in practically all species remain the same throughout the life of the insect, and usually even after its death. There are a few exceptions: some mature males acquire a pruinescence that may hide the normal body color, and in a very few species, most of them members of the grasshopper group, simple chameleon-type color changes have been observed. Where they occur, these changes consist mainly of a

A tropical weevil displays differing shades of green as his iridescent colors shift in the light.

The least bittern "blushes": the bare skin of its loreal patch between the eye and the bill may turn pink with excitement.

temporary darkening or lightening of the basic hue. The brilliant iridescent colors with which many insects are endowed and which are located in the scales that cover the wings, or in the surface layers of the integument, are based upon the same principles as those found in birds.

A look at some of the few exceptions among birds and mammals that have the ability to change rapidly the color of certain parts of their body shows that they always involve animals with bare skin patches, such as some apes, and birds with wattles or necks that are not feathered. When angry or excited, these skin patches may turn a vivid color, usually red or purplish. The most prominent example of this phenomenon is man himself, or, rather, those members of the human race that have light skin. In such individuals, color changes of the face brought about by a change in the emotional or physical state of an otherwise healthy person are often quite striking, and may reveal more about that individual's condition than his words. We know this from expressions such as "pink with embarrassment," "purple with rage," "white with fear," "blushing with shame," and "gray with exhaustion." In each case, the emotion causes a definite and noticeable color change in the affected person's face through an expansion or contraction of the blood vessels in the skin, and a proportionate increase or decrease in the flow of blood to that part of the anatomy. People with dark skin are naturally much less affected by such color changes, which are produced solely

24

by the amount of blood showing through the scantily pigmented epidermal layers of fair-skinned persons. Brown skin in such cases assumes a slightly more reddish or grayish tone, but the difference is not as noticeable as in an individual with a very light skin color, and is all but invisible in one with black skin. That does not mean that such persons do not blush, flush, or blanch, but only that all color-based evidence of these emotional changes is effectively concealed by the overlaying dark pigmentation of the skin.

Differing fundamentally from color changes in birds and mammals, the chameleon-type color changes involve altered pigmentation in the animals that display them, causing different hues and patterns to appear within periods ranging from seconds to hours or even days. In addition to the universally known chameleon, scores of other animals possess the unusual ability to alter their appearance in this way. They include many other lizards, frogs, toads, and fish, as well as invertebrate marine creatures such as octopuses, crabs, and shrimps. In view of the fact that the chameleon's fame is based entirely upon this ability, it seems strange that very few people are aware that so many other animals also possess it, some of them to an even larger degree than the chameleon.

Although the mechanism that activates the color changes differs widely with the species, being relatively simple in some and extremely complex in others, the basic ingredient is similar in all cases. This ingredient consists of pigment cells called chromatophores, and especially of those that are filled with dark melanin, the pigment responsible for all black, brown, and gray color in the animal world.

There is nothing unusual about pigment-containing cells or tissues, which are found in practically all animals, but the melanophores of the chameleon and others capable of color changes are very unusual indeed. Star- or blossom-shaped in most cases, they contain tiny granules of melanin either concentrated in a pinpoint area in the center, or dispersed throughout the entire cell. In some instances, these cells are

25

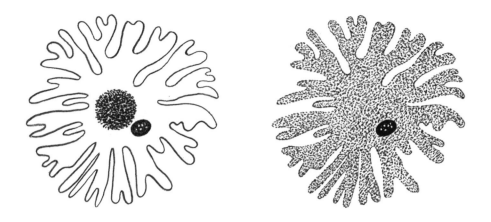

A typical melanophore: on the left, the pigment is aggregated in the center, and on the right, dispersed throughout the cell.

under the direct control of the animal's nervous system; in others, the endocrine system and its secretions influence the activity of the pigment granules. In still other species, both the secretions and the nerve impulses may play a part in activating and controlling the melanophores. In every case, however, the effect is the same: dispersal causes the animal to appear darker, and concentration makes it lighter colored or pale.

Although some animals among our quick-change group are capable of little more than such a darkening or lightening of their basic skin color, most have a much wider range, owing to a more complex and varied chromatophore arrangement. In addition to the melanophores, there may be present in the skin tissues of such animals lipophores, cells containing either red pigment (erythrophores) or yellow and orange pigment (xanthophores), as well as guanophores, cells filled with guanine crystals that act as light reflectors. If a cell combines granules of melanin with guanine crystals, it is known as a melanoiridophore. In a few animals, chromatophores containing several pigments, and therefore several colors, may be found. These are called allophores, and although rare, they are instrumental in creating some of the most striking color combinations found in the chameleon-type changes.

Although all the pigment cells have an important role in these displays of changing hues, the melanophores are by far the most active. They may play a dual role by covering or exposing other pigments and by creating, in combination with these other color substances, a wide variety of hues. Let us assume, for example, that an animal has a top layer of yellow xanthophores, a middle layer of melanophores, and a bottom layer of guanophores, or iridocytes, as those cells that produce mainly white or pearly hues are called. If the melanophores are "closed," i.e., their pigment concentrated in the center, light reaches the guanophores and is reflected by the crystals. Because very small particles sus-

Diagram of a color-change structure: yellow and black pigment cells and blue reflector cells combine to create different color combinations.

pended in a different medium reflect especially the shorter wave lengths, the blue light is reflected, but reaches the observer's eye only through the filter of yellow pigment cells. Accordingly, the animal now looks green. If, however, the blackish-brown pigment is fully dispersed throughout the melanophores, it obscures the guanophores and combines with the yellow of the top layer to make brown. A partial dispersal, permitting some blue light to come through, results in an olive or brownish green animal.

In this way, any number of color combinations can be achieved, depending upon the different pigments and their arrangement in the integument. We shall see in the following chapters how greatly this arrangement may vary in the individual species.

In addition to the physiological color changes occurring through aggregation and dispersion of pigment in the cells, scientists recognize morphological color changes. The latter are a rather slow, often semipermanent process that may take days or even weeks, and consists of an actual increase or decrease in the number of chromatophores. Morphological color changes may thus be compared to the tanning of a person's skin through long exposure to the sun, or, conversely, to the blanching of the skin occurring in people forced to stay indoors for a long time.

Morphological changes are almost always a matter of protective coloration, of adaptation to a particular environment or background. Physiological changes, on the other hand, may occur for a variety of reasons, some obscure, some obvious as an adaptation to a substrate, for instance, or as an emotional response.

One of the most startling and radical examples of morphological color change is provided by experimentation with the olm, a peculiar, almost sightless salamander that lives in a permanently larval state in the total darkness of deep caverns in southern Europe. The olms brought up from those lightless depths are almost pure white, with the exception only of the external gills, which are carmine red. If the ani-

The European olm showing adaptive color change: at the top, the salamander as it looks after it has been brought up from the depths of the caves in which it lives; at the bottom, the olm after it has been exposed to light for some length of time.

mals are kept in darkness, they will remain white, but if they are exposed to light over any length of time, they gradually turn dark until they are almost black. This proves that the skin of these creatures is very susceptible to light, and capable of forming melanophores when stimulated. A black olm returned to darkness soon regains its white or very light pinkish hue.

For more than a century, the physiological and biochemical mechanisms of color changes have been studied intensively by scientists. Although there still is much that we do not know about the intricacies of the process, we have in most cases found out what physiological factor or factors control the color changes in the animal's body, what part is played by outside influences, such as light, temperature, and humidity, and what kind of reactions result from artificial stimuli, such as injections of certain chemicals, or electrical shock.

We have had much less success in trying to answer the question of *why* these animals change color, especially in view of the fact that many of their closest relatives do not. In the days when many biologists believed that the concept of natural selection demanded an explanation on the basis of its functional "survival" value of every single phenomenon in the animal world, much was made of the fact that the color changes were adaptive, serving as a camouflage device. In the meantime, increasing attention has been paid to animal behavior in all its aspects, and careful observation has shown that the camouflage theory supplies, at the most, only one part of the answer. Concealment seems to be the primary function of color changes in *some* animals, and *some* color changes aid concealment in other instances. In the majority of cases, however, it appears that camouflage is only one of the functions fulfilled by these unusual changes in appearance, and in a number of species exhibiting very striking color changes, camouflage plays no part at all. Thus the question remains: Why and to what purpose do these animals change color?

At the present, considerable evidence points to the possibility of these changes serving primarily as an often complicated mode of self-expression, or "language," through which these animals communicate moods and emotions, as well as physiological conditions. In this context, the phenomenon becomes even more fascinating than before, especially since many of the animals involved are generally not considered to have such reactions as moods and emotions, which most people think of as being the exclusive monopoly of the higher animals. Getting to know more about the unique and often complex "color language" of reptiles and amphibians, fish and cephalopods may help to narrow just a little the tremendous gulf that has always separated man from these cold-blooded animals, and may permit us to see them as an interesting and valuable part of our natural environment. After all, these creatures, which preceded man by hundreds of millions of years, have senior rights to living space on this planet. Making sure that these rights, along with those of other wild animals, will be protected in the future as a service to man himself should be everybody's concern.

2. Dwarf Lions and Their Kin

LOOKING remarkably like a miniature arboreal triceratops, one of the oddest of the dinosaurs, the chameleon sits motionless, its peculiar feet half encircling the branch on which it perches, its prehensile tail coiled around a twig. Only the reptile's eyes move, swiveling in a semicircle as they scan the surroundings for a movement heralding either a prey or an enemy. Its color, a dull, mottled brown, makes it appear almost a part of the branch. Suddenly, the animal spots something in the foliage: a praying mantis on the branch below, which had been waiting for a locust to come within striking distance, has made its kill. But that small movement has betrayed the mantis to the sharp-eyed reptile. A long tongue shoots out, and in a blur of motion too fast for the eye, both the mantis and the grasshopper disappear into the chameleon's mouth.

Its hunger temporarily sated by this rather substantial meal, the lizard settles down to restful contemplation. But before long, its siesta is interrupted by the arrival of another miniature saurian of its own species. Outraged at such impudence, the first chameleon gets ready to make the intruder respect territorial rights. Puffing up its body and inflating its throat, the reptile rapidly changes from brown to a green color with a pattern of sharply delineated dark spots and light yellow lines. This threat display has the desired effect; the intruder backs off and retreats, and the first chameleon settles down again, slowly resuming its neutral brownish coloration.

Nearly 2,500 years ago, observers in ancient Greece were fascinated by a small tree reptile widely found in all the countries along the southern fringe of the Mediterranean, from Palestine to the Iberian peninsula. It may have been the slow, stealthy movements with which this creature stalks its prey, or it may have been the tawny color it so frequently displays that caused the Greeks to name the reptile *chameleon*, which literally means "dwarf lion." Aristotle, who gives a detailed description of the chameleon in his *Historia animalium*, does not tell us anything about the origin of its name, which has persisted throughout the ages, eventually becoming a byword in many languages. Because of this, the common chameleon is probably one of the best known of all small reptiles. The "common" part of its popular name refers to its extensive geographical range in the Old World as well as to its accessibility, for this is the only species that occurs in densely populated regions that have been cultivated by man for thousands of years. Through the centuries, it has been kept in captivity for display more often than any other species. Its scientific name *Chamaeleo chamaeleon* designates it as the prototype of an entire group. Almost all the other chameleons, some eighty species, are found in Africa south of the Sahara desert and on Madagascar. Only four occur farther east in Arabia and India. Most of these tropical chameleons are hardly known except to specialists, and some were discovered comparatively recently. One species from the Congo, for instance, was first described a few decades ago by the late Karl P. Schmidt, noted American herpetologist and former curator of the Chicago Museum of Natural History.

By any standard, chameleons are fascinating though rather ugly little beasts. They do not look at all like typical lizards. The average length including the tail is about a foot, although some are much smaller, and

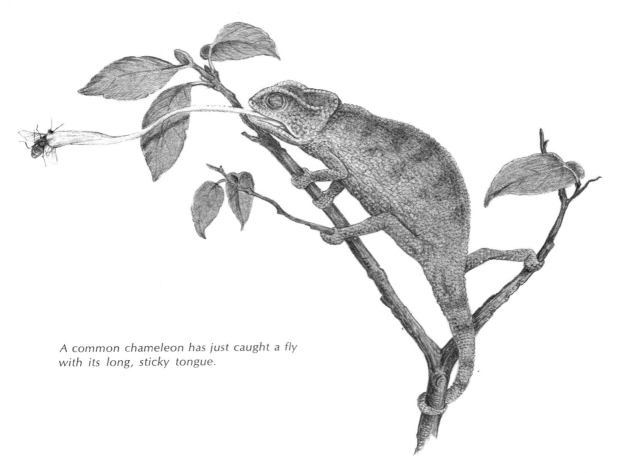

*A common chameleon has just caught a fly
with its long, sticky tongue.*

one, the giant of the group, measures almost twice that length. An outstanding feature of many species—though not of the common chameleon—is the bizarre adornments including helmet-shaped crests and hornlike projections worn especially—or exclusively—by the males. No one has so far come up with a satisfactory functional explanation of these appendages.

The chameleon's body is flattened sideways—a characteristic of many arboreal lizards—so that the head appears very wide when seen from the front. The large, bulging eyes are almost totally covered by circular eyelids that leave only small peepholes exposed in the center, giving the false impression that the chameleon cannot see very well, which is far from the truth. Its vision is in fact quite remarkable. The eyes can swivel upward, backward, and downward in a complete hemisphere, each eye moving independently of the other. Thus a chameleon may be looking at an observer with one eye, while watching, with the other eye, some movement occurring behind its back on the far side. Without turning its head or moving any part of its body except the eye, a chameleon is thus able to see everything going on in its immediate surroundings.

Scarcely less unusual is the reptile's tongue, which it employs like a spear gun in hunting its prey, except that the sharp point of the spear is replaced by an adhesive as a means of securing the victim. Among all reptiles, only chameleons use their tongue in this way, shooting it out at lightning speed, often to a distance that may exceed the combined length of its head and body, to capture an insect or other small animal. Three factors account for the ability of the tongue to function in this fashion: the extreme length to which this organ can be stretched; two sets of muscles, one to propel and one to pull back the tongue; and finally, the sticky substance with which the clublike tip is covered.

Another one of the chameleon's anatomical peculiarities is the construction of its feet. Instead of the characteristic reptilian foot with long toes, found in most other lizard families, chameleons have toes that are bunched together in opposing bundles on each foot, two toes on one side and three on the other, united except for the last joint that bears the claw. This arrangement, while depriving the chameleon of individually movable toes, turns each foot into a kind of strong grasping tongs. Once a chameleon gets a good hold on a branch, it is almost impossible to dislodge it except by brute force, especially because the animal's hold even with only two feet is reinforced by its prehensile tail.

It would seem that these distinctive features are enough to have warranted more than ordinary interest in this creature, but it was the chameleon's ability to change color that fascinated people more than

Common chameleon.

anything else, and that still holds true today. It usually comes as a surprise to most people when they learn that many other species of reptiles and other animals are capable of such color changes and that some have a much more extensive color repertory than our "dwarf lion," which can make no claim to the title of champion rapid-change artist. Most people are equally surprised to find out that chameleons do *not* change color just to fit their surroundings, and that attempts to in-

The brownish color of this common chameleon proclaims a fairly tranquil state. At right, a pattern indicative of excitement or fright is displayed.

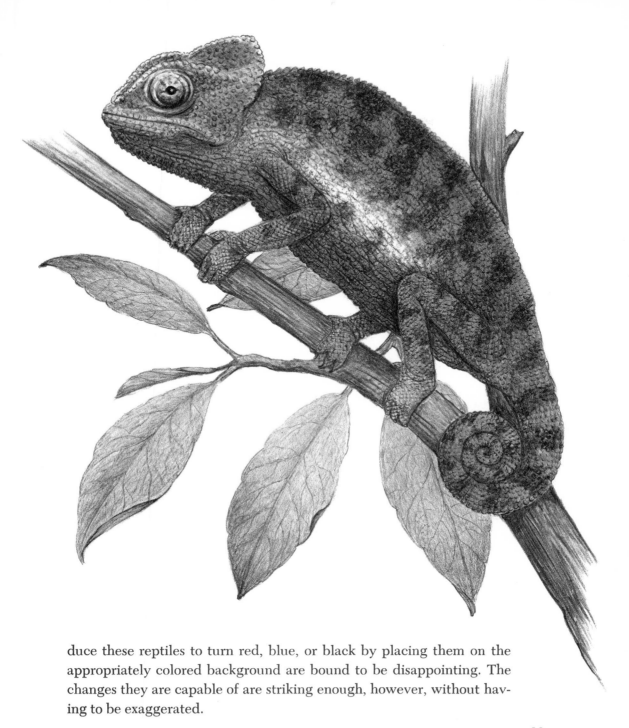

duce these reptiles to turn red, blue, or black by placing them on the appropriately colored background are bound to be disappointing. The changes they are capable of are striking enough, however, without having to be exaggerated.

In the morning, an undisturbed and placid chameleon usually displays a tawny color, with two irregular, reddish brown stripes along the length of its body on both sides, and a few dark spots or blotches visible on its back. When the animal is disturbed or captured during these hours, the spots become dark green while the stripes turn a light beige. The color of unmolested chameleons tends to change gradually as the day advances, becoming an overall grayish green or olive brown in which practically all pattern is obscured. A state of great excitement, on the other hand, will result in a general darkening and in the appearance of a sharply defined pattern with contrasting colors: on a dull green background the stripes show almost white and the spots are completely black.

It is evident from this color range that any of the various phases would most likely have some protective value in the trees where the chameleon normally lives. Adaptive coloration is an important factor in the chances for survival of many animals, especially harmless and defenseless creatures such as the chameleon, which is beset by numerous enemies. These include herons, storks, snakes, and even large lizards. An old proverb from southern France maintains that "a chameleon that is seen is a dead chameleon."

The changes in color, despite the fact that they reflect an emotional register ranging from fear to triumph as well as certain physiological conditions, usually do not impair the concealment value of the chameleon's color pattern. Usually, but not always: an excited or angry chameleon may display a green color where brown would be a better camouflage, and vice versa. This seems to be true regardless of the animal's color repertory. One of the common chameleon's relatives from the Congo, for instance, has a fairly extensive color range, including dark green, light yellowish green, yellow, white, and light reddish brown. When this chameleon fights with a relative, as frequently happens among these animals, it may turn an unwise brown in foliage where green would offer it better concealment. Despite such instances, these changes appear to permit—at least among chameleons and related species—a wide-range visual expression of emotional and physical states and still hold to a minimum the danger of making the animal look conspicuous. In some cases, the color changes may indeed add to the camouflage effect, although expression of moods and emotions appears to take precedence over concealment.

Within a matter of minutes, the rather pale colors of this African chameleon can be changed.

The pronounced pattern of this chameleon indicates excitement or fright.

Observations of chameleons and other lizards in captivity have given rise to the theory that some color changes have a defensive value unrelated to camouflage. Adolf Portmann, noted Swiss zoologist and author of many nature books, tells of a chameleon that was pursued by a dog. Turning around to face the dog, the chameleon at the same time

42

assumed a very dark, almost blackish color. The dog, apparently taken aback by the sudden change in the animal, discontinued its pursuit. Although additional evidence is needed, it seems entirely conceivable that a rapid and dramatic color change by a pursued animal could confuse and throw off balance a pursuer long enough for the intended victim to make its getaway.

Some naturalists have come to the interesting conclusion that the peculiar, measured, jerky movements with which chameleons stalk their prey may be as much an aid in camouflaging them as any of their color patterns. Because of their laterally flattened bodies, their slow, swaying progress along a branch very likely makes them appear, to some of their victims and enemies alike, as nothing more exciting than a leaf stirred by a breeze.

Of all reptiles, only the lizards possess the ability to change color rapidly. The two ancient groups of turtles and crocodiles are incapable of any changes; among snakes, color changes are limited to a slow, semipermanent darkening or lightening of the basic colors with which some species adapt to their environment. This has been observed especially among rattlesnakes, where very dark or very light substrates resulted in a gradual adjustment of the snakes' colors to those of their surroundings.

The large order of lizards has a number of families with considerable powers of rapid color change, but curiously enough, the most representative group within the order, the lacertids, lack all ability for such changes. Strictly an Old World family, they include such typical and handsome species as the fence lizard and the jeweled lacerta, but their colors are static, remaining the same at all times during the life of the adult animal, except for the dulling of their body patterns shortly before the periodical shedding of the skin, and the brightening of the hues after the molt.

In addition to the chameleons, the ability for rapid color change is found mainly in three major lizard groups: the geckos, agamids, and iguanids. All of them have members capable of such changes, although there exists great diversity not only in the range of hues that can be displayed by the different species, but also in the speed with which the individual can assume another hue. Some, like the North American horned toad or collared lizard, are capable of little more than a darkening or lightening of their normal patterns under the influence of light; others have color repertories that rival those of the chameleons.

Color changes among geckos, which occur throughout the warmer parts of the globe, are by and large less varied than in other lizard groups. The reason is that most geckos are active mainly at night, and have the drab, dull colors typical of most nocturnal creatures. Some geckos, however, are masters of camouflage, with colors and patterns so closely resembling objects in their natural environment—lichen-covered

Low temperatures and bright light have a darkening effect upon the collared lizard; these changes may, however, also be caused by emotional factors.

High temperatures and darkness may pale the western collared lizard to the color of sand.

branches, for example—that even the trained eye of a naturalist will, more often than not, fail to detect them as long as they do not move.

Color changes are chiefly found among those geckos that are active by day, and most often among those that are completely diurnal animals. An interesting group of brightly colored day geckos lives on Madagascar. These chiefly green lizards change their color especially while fighting among themselves.

A much larger color repertory, combined with a greater readiness to respond to different stimuli by color change, is typical of many species of the agamid family. No native agamids occur anywhere in the Western Hemisphere—the approximately three hundred members of this group are found in the Old World, with the greatest concentration by far in Asia and Australia.

Perhaps the most spectacular among the agamid species capable of color changes is a foot-long Indian lizard of the genus *Calotes*, popularly known as bloodsucker. This ominous name is most misleading, for the reptile is a harmless creature that confines its hunting to the various insects that share its habitat, and does not suck the blood of anyone. Its scientific species name *versicolor* is much more appropriate. The popular misnomer probably arose from observations of the reddish color assumed by angry or victorious males. Being quarrelsome, like most lizards, bloodsuckers readily engage in fights over territory and females, even though a good part of these battles usually consists of efforts to outbluff the adversary. The contests may be observed without disturbing the combatants—locally nicknamed "fighting cocks"—which at such times are more or less oblivious to their surroundings.

A battle between two male bloodsuckers starts with a ritual of preliminary shenanigans. They expand their dewlaps, erect their crests, and vigorously bob their heads up and down. Then the fighting begins in earnest, and both males go through a series of rapid color changes as they attack. Normally a kind of brownish or olive green, each now turns alternately lighter and darker, until one of them shows signs of weakening. When that happens, the stronger one presses his advantage, and final victory makes his head turn a bright blood red, while the rest of the body assumes a dull brick color. The loser, on the other hand, turns a drab grayish brown, a coloration very similar to that of the female.

Territorial fighting and the accompanying behavior, including changes in color, is not limited to the males. Females may fight vigorously to defend breeding territories against other females. In one such fight, a female of a close relative of the bloodsucker was observed angrily bobbing her head at the intruding female, who sat below her on the ground. The defending female was dark brown, her small dewlap almost black. The other female showed a pattern of turquoise spots on a background of leaf green. When the observer terminated the fight by capturing the dark female, she immediately changed color in his hands, assuming the same shade of green as her opponent had displayed during the battle.

In many species, color changes are an important part of the courtship ritual that must be enacted before mating can take place. In some cases, observations have shown that the color changes signifying sexual excitement that accompany courtship behavior, as well as the behavior itself, are very similar to those that take place during the fighting between males, except that in courtship these displays are meant to impress the female instead of the rival male.

When a male bloodsucker spots a female, he turns sideways to exhibit the silhouette of his expanded dewlap and his crest. This helps to increase the overall impression of the male's size. He then changes from olive green to a light yellowish or tan color, with a conspicuous black spot on either side of his dewlap, and begins to bob up and down by bending and straightening his forelegs exactly as he would in a contest with another male.

Although the majority of color changes we have observed among agamid lizards were related to fighting over territory or to courtship rituals, patterns and hues are known to shift under the influence of other emotions. One naturalist reports that the Australian bearded lizard, ordinarily a dark olive color, turns bright yellow, barred with orange, whenever it feels threatened or cornered.

Despite the fact that observation and research in the field, in captivity, and in the laboratory have yielded much information about the factors contributing to the phenomenon of color changes, zoologists are still far from having all the answers. Some of the most extensive experiments have been conducted with the green anole, a common species of arboreal lizard from the southern parts of North America. These small, slender lizards belong to the iguanids, a family noted for the diversity of its members in size as well as in body shape and habits.

Iguanids are found almost exclusively in the Western Hemisphere.

Many of them are capable of color changes, and the little green anole *Anolis caroliniensis* is typical of those species that can change their color very rapidly and radically. Because of this ability, these small reptiles are sold as chameleons in many pet stores throughout the United States. Actually, anoles are not closely related to the chameleons, and do not resemble them in the least. Between five and six inches in length, of which more than half is taken up by the tail, very slender, and normally a beautiful, bright, golden green above and greenish white below, anoles are graceful, agile, fast-moving arboreal lizards. In contrast to chameleons, they capture their insect prey, not by lying in wait or slow stalking, but rather by active hunting, lightning fast strikes, and sometimes, by a sort of flying leap.

The green anole is one of the most popular reptilian pets in the United States. All the same, very few people know enough about their habits to give them proper care, with the result that thousands of these unfortunate little "chameleons" starve to death every year.

Just as in the case of the true chameleon, the anole's popularity is based mainly upon its ability to change color, and to do it very rapidly. Except for a light-colored irregular stripe that runs the length of the spine along the back in many adults, and a certain amount of dark mottling in some individuals, this small lizard has no body pattern to speak of. When it changes color, the bright green of its upper side may turn a grayish or brownish green, and finally, a medium brown. The underside, although its hue varies from greenish white over light gray to a pale brownish beige, always remains very light.

As in the case of the true chameleon, it is still widely believed that the only reason for the anole's color change is the need for camouflage, and that therefore it will turn brown when sitting on branches or on the ground, and green when resting on, or climbing around in, the foliage. A number of natural-history books picture the anole in such camouflage positions. However, extensive observation and experimentation have failed to yield any solid evidence to support the color-

Green anoles in brown and green color phases. Background color has little to do with the changes.

change-for-camouflage theory. On the other hand, it has been proven beyond doubt that the anole cannot be induced to turn green or brown simply by placing it on an appropriately colored background. One or more other conditions have to be fulfilled in order to get results. That some of these other factors take precedence over any camouflage effect is easily proved by experiments in which anoles turned brown in all-green surroundings and vice versa.

Two of the most important outside factors influencing these color changes are light and temperature. Normally, an anole placed on any color background but kept in a cool room at a temperature of about 50 degrees Fahrenheit will turn brown, regardless of the amount of light it receives. If the temperature is raised to 70 degrees, the anole will turn green, but only if the light is very dim. In bright light, it remains brown up to a temperature of 85 degrees. Any further increase in temperature results in the lizard's displaying an in-between shade of olive or grayish green, which it assumes regardless of the light intensity.

Although the findings of such experiments seem to apply to all green anoles—"green" in this case being the popular name of this species—of both sexes, these color changes occur only when the animals are in a fairly calm and tranquil state. Anoles that are fearful or excited react differently. Emotional factors, which quite probably are the strongest influence in some rapid color changes, may completely alter or even reverse the normal pattern of change, especially with so lively and temperamental an animal as the anole. My own numerous experiments with different green anoles of both sexes have convinced me that light and temperature effect the usual color change *only* if the mood of the animal is conducive to such a change.

What my experiments confirmed was the presence, during daylight hours, of two basic color phases, the activity coloration and the repose, or neutral, coloration. While bright green—the activity color—may appear in a number of different circumstances, it evidently is *always* displayed whenever the animal feels insecure, endangered, or uncomfortable, as well as during periods of exertion. Brown, the repose color, on the other hand, apparently means that the animal is comparatively calm, tranquil, or sleepy, regardless of the color of its background. Between these two extremes the anole may display a number of mottled green-brown or olive hues, indicating moderate activity.

While discussing the pros and cons of the camouflage theory we must keep in mind that, under normal conditions in their natural environment, most of the color phases available to these animals would have some, and often considerable, protective value. What stands in question is not whether the various shades lend more or less protection, but whether the need, in given surroundings, for camouflage and concealment controls and regulates the changes from one color to another. Evidence seems to rule out the latter supposition.

If, on the other hand, we accept the premise of basic activity and repose colorations, it is interesting to speculate that these colors may very well have a protective value, not so much because of the animal's background, but because of its own condition. A resting, motionless brown anole would tend to resemble a branch or twig, while an active green-colored anole climbing around in a tree or bush would be much more suggestive of a leaf that is stirred by a breeze. In both cases, a predator would be less likely to notice the animal.

In my most recent experiments involving two green anoles, the animals were nervous, restless, and bright green when I brought them home and transferred them to a terrarium. Their apprehension about their strange surroundings was apparent from their restless prowling and from their attempts to get out of the cage through the wire-mesh top. Another sure sign of their discomfort was their refusal to eat even choice tidbits such as live flies. They stayed bright green—and restless —for the better part of two days, regardless of the sudden and drastic changes in light and temperature to which I subjected them during this period. With temperatures ranging from a low of 50 degrees to a high of 78, and in light of various intensities from very dim to very bright, the anoles had ample opportunity to react to these environmental factors. However, the reaction set in only toward the end of the second day, by which time they had calmed down considerably, a fact demonstrated by their acceptance of several live flies. Exposed to bright light thereafter at a temperature of 72 degrees, the color of both anoles changed, one to all brown, the other to a mottled green brown. In what can be described only as a deliberate snub for the camouflage enthusiasts, the brown anole proceeded to snooze blissfully on a bright green leaf.

In order to test further the "activity color" theory, which I had conceived after earlier observations, I decided to interrupt rudely the brown anole's nap and chase it around the terrarium in a feigned capture attempt. If the idea had any merit at all, the anole would have to turn green fast during the chase. Looking at the clock to time the start of the experiment, I pushed the anole from its perch (with a mental apology, of course) and chased it from one corner to another. It took less than sixty seconds for the lizard to begin changing color: the legs turned green, and then the sides. I had already observed earlier that the eyelids were the last to turn brown and the first to turn green again in any color change.

As soon as the anole had become completely green—some three minutes after the start of the chase—I withdrew my hand, and was careful not to disturb the animal again for the next hour or so. Within fifteen minutes after the chase had stopped, the anole began to turn brown, and remained that color until I repeated the performance an hour later, with identical results. Further tests with the other anole substantiated my findings.

Continued close observation tended to confirm the theory that green is the normal activity color of this anole during the daytime, regardless of whether the activity consists of just prowling around, escaping from an enemy, or hunting and capturing prey. In addition, green may stand for a number of other emotional or physical states, such as sexual excitement, fear, discomfort, or illness. Some of these I discovered as results of deliberate tests, some quite accidentally. I found out, for example, that as soon as I took an anole in my hand, it would struggle mightily at first and turn green. At the same time, the heartbeat accelerated. If I continued to hold the animal, the green turned a bluish shade, and a "black eye" appeared, that is, a black area appeared just behind the eye, half encircling it. This I learned to associate with extreme fear or stress, especially when an anole that was sick and emaciated when I acquired it, and which remained a horrible shade of pale grayish green throughout the remaining few days of its life, developed these "black eyes" shortly before it died.

A happier experience was the opportunity to observe the courtship rituals of two of my anoles. I introduced a male into the terrarium inhabited by a female I had had for some time. He immediately became enamored of the demure little female, and started courting her on the very next day. She played it coy, though, a smart tactic that had him following her around, bobbing his head, puffing out his dewlap to show its red color, and exhibiting a bright golden green color in marked contrast to the rather dark brown of the female. In a scene not easily forgotten, she was resting comfortably in the crown of a small plastic tree, one eye cocked down at the poor fellow below who stood erect on his

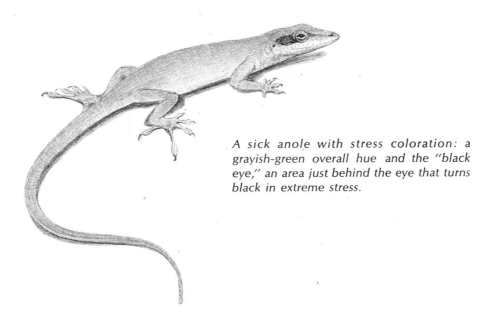

A sick anole with stress coloration: a grayish-green overall hue and the "black eye," an area just behind the eye that turns black in extreme stress.

Courtship behavior in green anoles: the scarlet hue of the male's expanded throat fan contrasts strongly with its overall golden green "excitement" coloring.

hind legs, grasping the thin trunk of the tree with his forelegs and bobbing his head vigorously while looking up at her.

I was watching this performance with a friend, and before I could get a word in about the balcony scene, she had already tagged the two: Romeo and Juliet. And that is what they have been called ever since.

It probably was the balcony scene that melted Juliet's heart and made her decide to accept Romeo; in any case, they mated several times. As throughout the entire courtship, the female remained brown during the mating act, except for a small piece of skin on her neck which the male grasped firmly between his jaws: that small area was bright green. The male was green at first, but turned brown later. This sequence of color change apparently occurred every time these lizards mated.

The healthy anole in its natural environment undoubtedly goes through a number of color changes in the course of a normal day as periods of activity alternate with periods of resting, and as environmental factors such as light and temperature exert their influence. To anyone willing to take the time and patience to acquaint himself with these reptiles' "color language," it soon becomes quite as intelligible as some of the more common and conventional kinds of animal communication.

There seems to be no doubt at all that periodical color changes in the green anole signify a state of health, well-being, and normalcy, whereas animals maintaining their green coloration unchanged over long periods of time indicate nervousness, apprehension, fear, or discomfort.

All these conclusions apply to daylight hours only. At night, every anole I ever observed was green. The nighttime green is usually somewhat lighter and not quite as intense as that displayed during the day, but it is a definite green. It is interesting that the activity color by day is similar to the sleep-time color of the night; this can be established easily enough by anyone willing to sacrifice a few nights of uninterrupted sleep for the sake of exact knowledge. Regardless of the tem-

Calotes mystaceus, an Indian agamid lizard, in its normal brown repose and low activity coloration.

perature, I have never found any evidence of even the slightest color change at any time during the night.

The fact that a nighttime resting color or color pattern exists among at least some reptiles capable of rapid color changes is confirmed by observation of different species. I found evidence of it in experiments with a lizard of the genus *Calotes,* a close relative of the aforementioned "bloodsucker" of India. The animal was evidently badly shaken by the long journey from its native habitat and the unfamiliar and uncomfortable conditions it had been subjected to during the trip. In any case, when I acquired it, the lizard was painfully thin and completely lethargic. With half-closed eyes, its body limp and unresponsive, it permitted itself to be picked up and handled without struggling. The color was a grayish, mottled brown, with a very indistinct pattern of some lighter lines and reddish brown spots running along the back on both sides of the spine. Food was ignored, and so was everything else; without moving, the lizard sat on, or rather clung to, a branch in the terrarium, looking very much like a part of the wood.

Assuming that the temperature in the room, a comfortable 72 degrees Fahrenheit, was still very low for an exotic creature, I placed the terrarium directly above a warm radiator and in a position that exposed one half of the cage to the bright winter sun. Within half an hour the lizard showed signs of animation. Its eyes were fully open, and it had climbed to the top of the branch. The most startling difference, however, was the change in color. On the head, back, and sides, the brown had lightened to a rather pale grayish green. This color was strongest on the animal's small crest, which looked like jade. The throat, earlier a mottled light beige and dark gray, had turned deep blue. The body pattern, so indistinct before, suddenly was clearly defined: a broad, light beige stripe ran all the way from the tip of the nose beneath the eyes through the ear opening and along the back, where it turned into a chainlike pattern linking the reddish brown spots, which alone had retained their original color.

Higher temperatures and increased activity brings about a change from brown to jade green in Calotes mystaceus.

Quite accidentally, I brought about a similar but much more radical and dramatic change when I became concerned about the lizard's continued refusal to eat for several days after I had acquired it. Fearing that it would become too weak, I decided upon an attempt to force-feed it. Holding it in my left hand, I gently pried open its jaws with my right, and tried pushing a mealworm into its mouth. It was not an easy job, for the lizard struggled mightily in my hand, but I finally succeeded in getting him to chew and swallow the insect. At that point, I noticed that the animal's head had changed color. Releasing it for a better look, I witnessed the fascinating transformation of a reptilian Cinderella. Except for the tail and the hind legs, every vestige of brown or gray had disappeared. Instead, the lizard now was a bright turquoise blue green on the head, neck, and crest, shading into jade green on the back and the sides. On this background, a pattern of cream-colored stripes and reddish brown spots stood out like a gem-studded necklace. The throat patch had turned a deep violet with purplish overtones.

I repeated the procedure next day, and was gratified to find that the results were identical. This time, I clocked the transformation, and found that the color change began exactly thirty seconds after I took the lizard in my hand, and reached its peak one minute and twenty-five seconds later. After that, I released the animal and the green color began to recede. This very rapid rate of change worked both ways— it took the *Calotes* only a little over one minute to become brown again.

Undoubtedy, these transformations were other instances of repose, activity, and stress colorations, even though the changes in this lizard did not seem to occur so readily and so frequently as in the case of the green anoles. However, one always has to keep in mind that all observations of animals whose natural environment cannot be exactly duplicated in captivity, or of those that are confined or restrained during an experiment, must remain fragmentary, and may sometimes be actually misleading. A case in point is the story told by Konrad Lorenz, famed German authority on animal behavior, about one of the dogs which Ivan Pavlov, the widely acclaimed Russian scientist, had trained to display a conditioned reflex in response to certain stimuli. When an American student released the dog from the harness in which it had been confined during the experiment, the dog reacted quite differently to the same stimuli, proving that the conditioned reflex was only a small portion of the total pattern. Pavlov became quite unscientifically furious and severely reprimanded the student. Among other things, this story illustrates the dangers of relying exclusively on—and reading too much into—experiments in animal behavior that introduce artificial conditions and do not permit the animal to be a free agent.

From observations of various *Calotes* species in the field, it seems certain that these lizards in their habitat go through numerous color changes during the course of a day in response to different moods as well as to environmental factors. From my limited experience with one individual it seems likely that the brown coloring of this species is typical of a state of repose or of low activity, which in turn may be partly due to the relatively low room temperature of 72 degrees Fahrenheit. Variations in the shade of brown, which may range from a very grayish tone to a much warmer and more yellowish hue, seem to indicate the difference between lethargy and contented repose. An acceleration of the animal's metabolism through activity, excitement, or exertion, or through higher temperatures is expressed by a change from brown to various hues of green, and is usually heralded by the color of the throat, which turns blue long before the head begins to assume its greenish hue.

As in the case of the sick anole, I was able quite by accident to ascertain that the display of the blue green coloring for long periods of time by the *Calotes* lizard signals a state of ill health. Possibly as the result of poisoning by insecticide residues, the lizard developed spasms and and partial paralysis especially of the hind legs. During the five weeks of this condition, the animal ate nothing and took only water. At the same time, its color remained greenish blue without any change day or

High excitement turns Calotes mystaceus *into a turquoise-colored lizard; the expanded dewlap displays a deep, purplish violet hue.*

night throughout the five weeks. When the blue finally started to change back to brown, I was fairly sure that the animal would recover, and it came as no surprise when the paralysis disappeared and the lizard began to take food once more.

Under normal circumstances, the *Calotes* lizard also has a distinct nighttime coloration. Unlike the anole, its daytime activity, or stress color does not double as the repose color at night. Not even the slightest tinge of green shows at night; instead, the lizard appears drained of all color, a pale, ashen gray that I never observed during the day. On this background, the stripes stand out almost white, and the reddish brown spots on the back are the only specks of color. The most interesting fact about this coloration is that the lizard becomes very conspicuous on its dark brown branch if a light shines on it during the night. Normally, of course, this would not happen in the animal's natural environment.

It would be fascinating indeed to keep a number of these lizards of both sexes in surroundings closely simulating their natural habitat, and observe their entire range of color changes as different moods, emotions, and physiological conditions are expressed in this way. From observations in the field and in captivity, a limited "color vocabulary" can even now be assembled. Grayish brown stands for great lethargy or defeat; a warmer brown for repose and minimal activity; grayish or olive green for normal activity; the same color with a blue throat for increased activity; bright blue green and a purplish throat for high excitement, exertion, and stress; and finally, pale ashen gray for nighttime repose.

The stimuli that trigger color changes in reptiles are predominantly visual, although the skin also has sensors that respond to light and to touch. This means that even a blinded animal is still capable of color changes, although on a reduced scale. In most species, the changes seem to be almost exclusively under direct nervous control; in some, hormonal activity plays a large part. There is no agreement among scientists on the influence of endocrine secretions on the chameleon's

The nighttime coloration of Calotes is so pale that the animal becomes conspicuous on a brown branch.

color changes; other lizards, on the other hand, have been proved to lighten or darken after injections with various hormones.

Chameleons have a chromatophore system that is much more complicated than those of most of its color-changing relatives. They have yellow, black, and red pigment cells, as well as deep layers of guanophores that act as reflectors. In the green anole, the arrangement is simpler. Directly under the skin, a layer of what looks like small droplets of yellow pigment fill the spaces between the larger cells. The second layer consists of cube-shaped guanophores filled with purine crystals. Finally, a deep layer of melanophores stretches below the structural cells, but its branches reach right up into the skin.

Both the melanophores and the light-refracting guanophores have a very important role in the changes from brown to green and vice versa. The fact that some of these hues are partly structural is best illustrated by a green anole in its various color phases. Depending upon the angle of the light source, the brown may appear grayish olive, medium brown, or coppery reddish brown. The green is less sensitive to changes in light angle, but it does shift its hue from a brilliant emerald color to

Reptilian chromatophores.

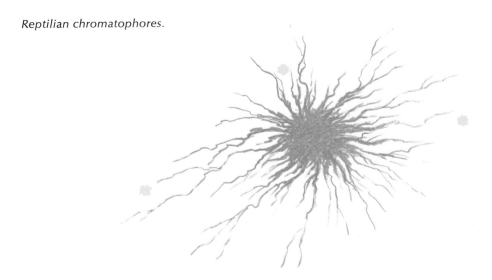

A diagram showing the action of the melanophores in the skin of the anole. Top: green color phase, melanin concentrated within cells. Bottom: brown color phase, melanin dispersed.

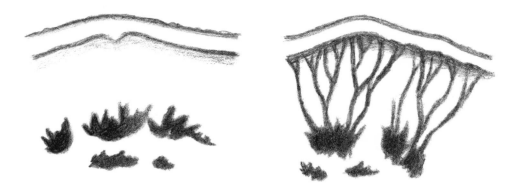

An electron micrograph of melanophore changes in the skin of an anole.

a duller green if the light source is moved or the observer's position altered.

Structure and shape of the scales vary considerably among the different species of lizards. Anoles have small pebble-shaped scales, whereas those of the *Calotes* lizard are much larger, sharply keeled, and arranged in overlapping rows. The brown color of these reptiles is not affected by the light angle, but the change in the green is startling. I have held this lizard during its green phase, slowly moving my hand to let the light come from different angles, and watched its bright turquoise color turn to a deep blue at an oblique angle.

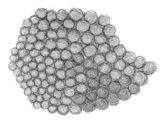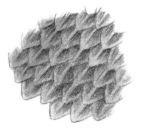

Small areas of scales of the green anole and the Calotes lizard, enlarged. In both cases, a color change from brown to green is in process.

In all the lizards, the yellow pigment cells seem to be relatively static. For both the brown and the green colorations, these cells have to be in a state of dispersal. Only the very pale, grayish color phases would seem to indicate aggregation of the yellow pigment; in all the others, the melanophores are the "active" cells, determining largely the amount both of pigment yellow and of structural blue that reaches the eye of the observer.

More fascinating even than the complex physiological mechanism of this "color language" is its code which, once it is better understood, can tell us a great deal about these animals, their moods, emotional responses, and reactions to changes in the environment. A lizard that lets you know when it is angry, excited, contented, or triumphant seems somehow much less "cold-blooded" than the average reptile. Further study of this language will undoubtedly add to the age-old fascination of chameleons as well as to that of their quick-change kin.

3. The Frog That Changed for Dinner

HUDDLED in a corner of the terrarium, its light-colored underside hidden from view, its golden-rimmed eyes half closed, the frog is almost invisible, so closely does its color match the blackish olive hue of the sphagnum moss on which it is resting. Even after a big fly is introduced into the terrarium, the frog remains motionless for a few moments, except for its eyes, which are now wide open. But as soon as the fly begins to buzz around, the frog straightens up, lifts its head, and turns to follow the flight path of the insect as though measuring the distance. Finally, there is a single, accurately timed leap, and a successful landing with the captured insect.

While the now wide-awake frog catches one fly after another, almost as quickly as they are placed into the terrarium, its color undergoes a startling change. The blackish hue disappears, paling to a light olive on the hind legs, and to grass green on head and back. The frog now resembles the conventional illustrations of its species, the common North American green frog, found in nature guides and other books on amphibians. Hardly anyone but an expert would have recognized the almost blackish frog of just a few minutes earlier as a member of this familiar and widespread species.

Although amphibians have by and large fared a good deal better than reptiles in their relationship with man, surprisingly few people know much about the habits and life histories—including the ability to change color—even of the more common species. Yet at least one large amphibian group—the frogs—has traditionally evoked tolerant and even friendly responses in man: the folklore of many nations features frog kings and princes in usually benevolent roles. Although closely related to frogs, toads have not been treated so well; these harmless creatures have been symbols of evil for centuries in many fables and superstitions, probably because of their generally dull coloring and habit of hiding in crevices and damp places. By comparison, a frog sunning itself on a lily pad or climbing around in a tree appears a happy creature of sunlight and open spaces. Even today, in our mechanized world stripped of myth and magic, a chorus of spring peepers in April inevitably stirs, in all but the most blasé, a feeling of gladness at the newly awakening life, for the voices of these tiny frogs are just as much a joyful sign of spring as the first robin or the golden blossoms of a forsythia bush. To anyone who grew up in the country, the sound recalls memories of collecting tiny polliwogs in glass containers, where they could be observed for weeks until the long-tailed, fishlike creatures, in a microcosmic replay of one evolutionary phase, gradually became transformed into tailless, four-legged frogs.

Amphibians are principally divided into two major groups: one comprising the tailless species such as toads and frogs, and the other, the tailed salamanders, which includes both the terrestrial and the semi-aquatic species that have no powers of rapid color change, and hence do not concern us here. Frogs and toads, on the other hand, are often accomplished color-change artists, and some display a wide range of different hues and patterns to suit different occasions.

Although frogs and toads are closely related and the points of difference between them are by no means clearly defined in all cases, a comparative sketch of some of their distinguishing features will serve to

A European salamander.

differentiate the more familiar species in the temperate zones. By and large, toads are broader and less streamlined than frogs. Having stouter bodies and shorter legs, toads are slower and cannot jump as well. Their skin is usually rougher and much wartier, and their colors are more often limited to the gray-brown-olive range—the hues that match those of the dead leaves and stones in the moist places in which they live. Only the undersides of some toads show bright colors, such as orange and yellow.

None of the amphibians have a poisonous bite, as do the venomous snakes. Many species have only in the upper jaw teeth that are useful

A frog (left) and a toad.

South American arrow poison frog
The brilliant coloring is thought by
some to be a warning.

for holding captured prey, though too small to inflict wounds. However, the skin secretions of some species, primarily toads, are toxic and can be extremely irritating if they come in contact with the delicate tissues of the eyes or mouth. And there are a number of pretty, colorful South American frogs whose skin yields a substance from which Indians make an arrow poison potent enough to paralyze even large animals within a short period. The exceptionally bright color patterns of these frogs are probably warning colorations advertising their inedibility, and all available evidence seems to indicate that few predators will attack them. The majority of amphibians, however, are nontoxic and quite harmless, their defense consisting mainly of their ability to hide and camouflage themselves. Some species, mainly toads, are beneficial because their food preferences include many insect pests as well as such unwelcome plant

72

feeders as slugs. In the temperate zones, even those amphibians with toxic secretions are innocuous if care is exercised to avoid contact with sensitive tissues.

Color changes among frogs and toads, long known to both amateur and professional naturalists, have been investigated in detailed studies only in recent years. We now know a great deal about the process involved in these changes—their internal chemical and physiological mechanisms, and the various environmental factors, such as light, temperature, humidity, or touch, that may set off a color change under certain conditions. Much less is known about the relationship between emotional factors and the changes in skin color, because relatively few in-depth studies of living animals in either natural or controlled artificial surroundings have been made.

The chromatophores in frogs are very similar to those found in reptiles, differing mainly in the manner in which the pigment in the cells is stimulated either to aggregate or to disperse. As we have seen, the chromatophores in reptiles are in the majority of cases under direct nervous control, which permits extremely rapid changes. In frogs and toads, color changes are primarily regulated by hormones discharged into the bloodstream, which activate the cells at a slower rate. Nevertheless, certain species of tree frogs can manage to put on a new livery within minutes, and some changes caused by emotional factors—fright, excitement, or fear—are quite rapid.

The frog's eyes are at the top of the list of organs that set off the chain of reactions resulting in a color change. Perceiving the color of its background with those marvelous eyes, the frog does its best to adapt to, and blend into, the surroundings—always assuming, of course, that the animal is undisturbed and reasonably tranquil. An excited, aggressive, or frightened frog may react quite differently.

The ability of frogs and toads to assume the general color scheme of different environments can be utterly confusing to a layman attempting to identify the various species. One naturalist tells the story of the

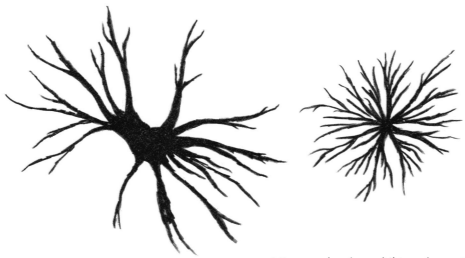

Micrograph of amphibian chromatophores.

elderly lady who for a long time was convinced that her small garden was inhabited by three different kinds of toad. One was a light green toad living among her rhododendron bushes; another, a spotted brown that made its home among rocks and dead leaves; and the third, an olive-colored toad she found at the bottom of her small pool. After many months, she finally realized, after observing the animal in each habitat more closely, that all three "species" were actually a single toad, clad in three distinct garbs at different times to suit different environmental requirements.

Although the chief stimuli for color changes in amphibians are visual, light sensors in the skin also play a role, as proven by sightless frogs that change color under the influence of light. In addition, the skin of a frog may be reacting to cold, warmth, touch, or moisture when changing color. Many frogs, for example, display different hues while sitting on a rough surface than on a smooth one, even if both surfaces happen to be the same color. Moisture may have a great influence on the color of some frogs, while others do not seem to respond to it at all. In one species, exposure to dampness was found to bring about a reversal of the usual adaptive color change. In this experiment, the frog darkened on a white background, because sensors in the skin, reacting to the moisture in the substrate, caused a dispersal of the melanin; extreme dryness, on the other hand, brought about a paling of the same frog on a black background. In this respect, the color changes in amphibians differ considerably from those of reptiles, which do not have this type of color response to tactile stimuli.

Prolonged exposure to certain background colors may cause semi-permanent, morphological color changes in which both the number of the melanophores and the quantity of pigment granules in the cells are either increased or decreased. This means a frog or toad may adjust to

The European green toad in two different color phases.

a very light or a very dark environment by assuming a more or less permanent light or dark color phase which, although reversible, would require a considerable period of time to change. Adaptations of this type are a common phenomenon in most animals capable of color changes.

Excitement and other emotional factors generally lead to a paling of the colors in frogs. But to make things a little more interesting, the South African clawed toad tends to darken when it gets excited. So does the peculiar Javanese "flying" frog, which is able to make great flying leaps with the aid of membranes stretched between its fore and hind legs to form a kind of parachute when unfolded.

Among all the amphibious groups capable of color changes, the tree frogs are the undisputed masters of the art. Though more closely related to toads than the so-called true frogs, tree frogs are generally much smaller than the other members of the clan, and are often brightly colored. This group is renowned for its powers of color change, which in some cases rival or even surpass those of the legendary chameleon. Tree frogs are also distinguished from other species by the sticky pads on their toes that permit them to cling to smooth surfaces and even to climb up the wall of a glass tank without any difficulty.

One of the most common tree frogs of Europe is *Hyla arborea*, a two-inch species that is normally bright leaf green above, light gray below, with a darkish brown stripe running from the tip of its nose through the eye along each side of the body.

This tree frog has always enjoyed great popularity as a pet for children, who consider it a good weather prophet and check its behavior daily—especially before outings—to determine whether it will rain. If the tree frog climbed to the top of the small ladder or branch in its cage and was silent, the weather could be expected to be fair. But if it came down and stayed in the water or on the wet moss at the bottom, and sounded off with its resonant croaks, beware! It was supposedly a sure sign that bad weather was coming. Those who smile at

The European tree frog in its normal coloring.

the idea of a frog metereologist should be reminded that despite all their sophisticated electronic equipment, scientists still have no instrument that heralds the coming of a hurricane as accurately as the behavior of a small Central American crab that scurries inland to escape flooding tides. And so our hurricane watch includes watching the crab.

I recall from my own experiences with pet tree frogs that they were far from infallible as weather prophets—although possibly I failed to interpret their "weather information code" correctly. On the other hand, it seems in retrospect that they were probably more reliable than our scientific weather reports of today. These considerations aside, watching them gave me great pleasure, especially because of their ability to change color, a spectacle that was endlessly fascinating to all of us tree-frog keepers, adding considerably to the appeal of these little animals.

Hyla arborea is not one of those tree frogs possessing a wide color range. The hue of green may vary a good deal with both the frog's

European tree frog in three different hues. The changes are mainly influenced by humidity and temperature and are often independent of the animal's background.

surroundings and its conditions. I remember from my observations that it could range from a very dark green to a yellowish, almost chartreuse hue. They usually turned dark green when sitting on the ground in the moss, and light leaf green on the branch among the leaves. The days before molting were the only times during which the color deviated from the green part of the spectrum; on those occasions, the frog turned a grayish blue or slate color, which was followed by an especially bright green shortly after the molt. This kind of color change, however, is not confined to animals that change their hues by pigment aggregation or dispersal; practically any periodical shedding of the skin in reptiles and amphibians is distinguished by a dulling of the colors before the molt, and a subsequent display of especially bright fresh colors.

I learned to interpret a paling of *Hyla arborea*'s color as a sign of excitement, discomfort, or fear, for whenever the frog was disturbed, or taken out of the terrarium, its color would lighten considerably, sometimes assuming an almost yellowish green hue, which quickly returned to the normal leaf green once the frog had again become comfortably settled. Moisture definitely had a darkening effect: the frog was always dark green when in the water or on wet sand or moss, as well as on rainy days.

One of the most common North American tree frogs is *Hyla versicolor*, a one-and-a-half-inch-long species that fully justifies its Latin species name by its astonishing capacity for color change. This frog may resemble a piece of lichen-covered bark one minute, and a little later change to bright green on a leaf, or to brown on a branch, or to gray on a stone. Along with its color, it can alter its pattern, becoming mottled or solid colored as the occasion demands. Most other tree frogs of the temperate zones are somewhat more limited in their color range. Some, like the North American green tree frog or the squirrel tree frog,

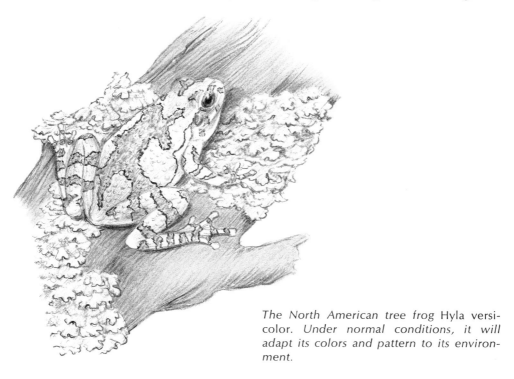

The North American tree frog Hyla versicolor. *Under normal conditions, it will adapt its colors and pattern to its environment.*

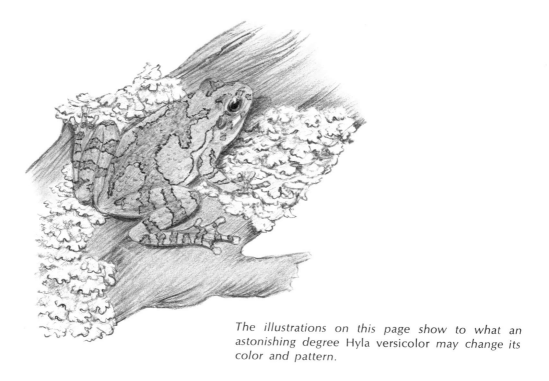

The illustrations on this page show to what an astonishing degree Hyla versicolor *may change its color and pattern.*

have little more than the ability to lighten or darken their basic hue; others, like the tiny one-inch spring peeper, may turn brown or green depending upon their surroundings. The most proficient color-change

The color changes of the spring peeper are limited to tan, brown, and green. This tiny tree frog is more often found in a brownish color phase.

artists are found in the subtropical and tropical regions of the world. Many South American tree frogs are relatively little-known, but captive specimens of some of these tropical species have displayed a remarkable color repertory in their changes.

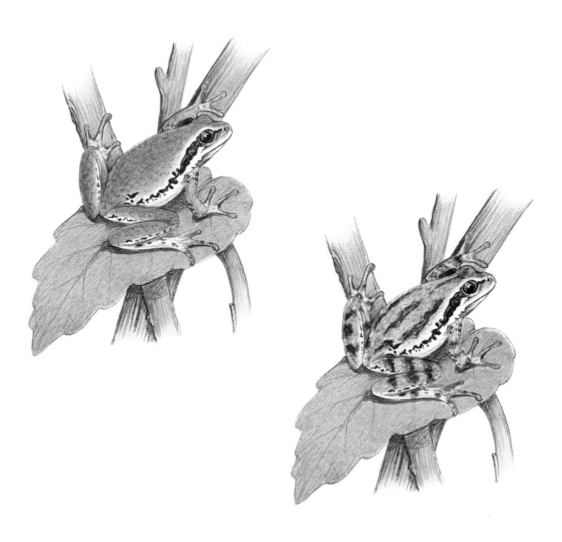

A very versatile species whose color changes have been closely studied is *Hyla goughi,* a tree frog found in Trinidad. This normally brownish member of the group can assume a broad spectrum of hues—from a very dark chocolate brown to maroon, reddish brown, orange, various shades of yellow, and finally, a very pale pearl gray. *Hyla goughi* is apparently surpassed only by a tree frog from West Africa

The Pacific tree frog is an accomplished color-change artist. It can replace a dark brown or mottled phase (left) with an overall dark green color in less than ten minutes.

whose color repertory has been pictured as quite unbelievable to any except those who have seen it. The late naturalist Ivan T. Sanderson described this frog as normally mottled green and brown on the head and back. After he picked up one specimen, it became almost black except for a terra-cotta-colored stripe along the back. Subsequently, he observed other individuals of this species exhibit an incredibly wide range of colors, including "pure white, yellow, gold, orange, brick, various browns, maroon, purple, mauve, pink, sea-green, grass green, and dove gray"—a record that no other known amphibian can even approach, and one that the most renowned among the reptilian color-change artists would be hard put to match.

The so-called true frogs, which have generally smooth skin, narrow bodies, long hind legs, and no sticky pads on their toes, are much less versatile in color change than the tree frogs. None of them can come close to displays such as those of *Hyla goughi*, or even *Hyla versicolor*. Members of this group in North America include the common bull frog and the green frog, as well as the attractively patterned leopard and pickerel frogs. All of these are capable of changing their basic color to lighter and darker hues, depending upon their condition as well as on the environment. Many of these frogs have been used extensively in laboratory experiments investigating the physiology of color changes. During these experiments, injections of adrenalin and other hormones were used to prove that the color changes in frogs are ac-

Another versatile artist is the Cuban tree frog, a relatively large species that can change its color within minutes.

A leopard frog a typical member of the so-called true frogs, as distinguished from the tree frogs.

tivated by hormones, and are not under direct nervous control, as in many reptiles. Relatively few live animals of this group have been observed over long periods of time in studies of the relationship of their behavior to patterns of color change under controlled conditions and in different environments.

During the past few years, I have kept a number of frogs and toads in terraria, partly because I enjoy observing small wild animal life of this type, partly because I wanted to make a detailed study of the color changes that had fascinated me even as a child. One of my star performers has been a green frog named Max—one of two frogs I received as a present two years ago from a friend who had caught Max and a smaller female in a pond on his property. Max has been with us ever since, and has become very much a member of the family.

Anyone who doubts that great variations in disposition and temperament exist among different individuals of the same species, even among groups commonly relegated to the lower classes of the animal kingdom, would have been converted easily just by observing these two frogs. From the beginning, it was quite obvious that Max had a broadminded, tolerant approach to people, and although he obviously considered them a mighty peculiar species, he had no serious objections to his captors as long as he was housed and fed well, and received enough attention. The little female—who remained nameless because she did not stay long enough for me to think of a suitable name— wanted no part of any relationship with humans. From the first hour, all she wanted was to get away, and she tried everything to that end day and night; after a few days, I finally gave her the chance to realize her ambition, mainly because I do not like to keep animals that feel so obviously uncomfortable in captivity.

Despite the fact that he had the same chance to escape, Max stayed behind, probably because he thought it worthwhile to give the am- phibian-human relationship a try. He settled down in his terrarium and has up to now given every appearance of being well satisfied with room and board. Whereas the female had stubbornly refused food during her untiring attempts to get out of the terrarium, Max accepted almost anything I offered him, although he was—and is—especially grateful for such choice tidbits as large flies, and learned to eat pieces of filet of sole only after some considerable coaxing.

Both frogs at first displayed a somewhat brighter green hue than the one I had observed in other members of this species. The female never once changed color in the few days she stayed with us, display- ing the same fairly bright medium green, especially on the head and back, while the hind legs were more olive colored. I later came to asso- ciate this color with one of two situations: either as a background adaptation to grass and other green plants of a medium hue, or as a sign of excitement or discomfort. The female quite obviously was not grass green for adaptive reasons: I had covered the floor of their tem- porary terrarium partly with gravel, and partly with very dark, almost blackish olive sphagnum moss. Max also remained grass green through- out the first day, but by the second day he had already settled down and displayed a much darker shade. From that point on the two frogs differed in color as well as in behavior. Max would change to a lighter, brighter green on a dark background only after he became excited and active in the pursuit and capture of prey, as described in the opening paragraphs of this chapter. Otherwise, he obliged me by changing his hue quite frequently when he was transferred to different-colored surroundings, most often adapting as much as possible to the general hue of the background on which he was placed. Almost invariably, he turned bright grass green whenever I put him outdoors on the lawn in a wire-mesh enclosure. At such times, it was quite astonishing to see how closely the color of the frog matched that of the grass, and how perfectly the animal blended into the background.

When artificial backgrounds were substituted for the natural ones, the results were much less clear cut. It was obvious that factors other than background coloration were exerting their influence. I once kept Max for days in a container lined only with a light yellow paper towel. He seemed quite tranquil in these strange surroundings, accepting food as usual, but remained dark green. During the winter months, despite the normal background of light green moss in his terrarium, room temperatures that were kept at an even 72 degrees, and regular feedings, he turned very dark and stayed blackish green until spring. I came to the conclusion that the shortness of the daylight hours and the consequent reduction of the amount of light in winter were at least par-

Max, a placid, companionable green frog.

tially responsible for this phase. The fact that his color began to lighten —without any change in environment—as the days became longer seemed to support this theory.

In the water, Max almost always displays a medium green hue, neither very light nor very dark, but never quite as bright as when he is sitting in the grass. The color of the water basin seems to make no difference at all: I have tried white, yellow, green, and black basins. He accepted them all without hesitation, and none of them seemed

After the first few days, Max began to turn bright green in grass, and dark on dark moss.

During wintertime, Max stayed dark even on a light-colored background.

to have any influence at all on his coloring. It would appear that moisture in this case is the controlling factor.

Max has a strong, resonant voice, and likes to use it. He sounds off whenever he feels like it, and can be quite persistent until he gets an answer. When he does receive an adequate response, however, he expresses his satisfaction with a few soft "ughs." He is able to tell one voice from another, and regularly waits until I come down in the morning: as soon as he hears me speak, he greets me with a few loud and cheerful croaks. A sociable frog, he likes company and takes part in conversations whenever I have guests. Fortunately, I have the testimony of several objective witnesses to Max's extraordinary behavior, or I would be afraid of being accused of telling tall tales.

Perhaps the most interesting fact about Max's voice in the context of this book is the relationship between the sounds he makes and the color he displays. He never "talks" when he is in a dark color phase and rarely when he is light green. He is usually most vociferous when sitting in the water or on wet moss and grass, dressed in his normal medium green garb.

In the beginning Max used to change color whenever I took him out of the terrarium and held him in my hand or let him sit on my knee.

Regardless of whether he had been dark or medium green before, he always paled somewhat on such occasions. By the time he was accustomed to my handling him and seemed to enjoy sitting quietly on my hand for prolonged periods, these color changes became less frequent, and finally rarely took place at all. This surprised me somewhat because my hand was a dry surface compared to the moss of his terrarium, and I would have expected a change in hue caused by that difference alone. Judging from his behavior, the initial color changes must have had an emotional, rather than an environmental, origin. I do not believe that handling ever really frightened him, but in the beginning he was reacting to an unusual experience. When it became a routine matter, it evidently no longer merited a color change. Of the many frogs and toads I have kept over the years, Max is the only one that has responded in this manner; all the others went through a color change—only slight in some, more pronounced in others—whenever they were handled for more than a few minutes.

Max had been with us for more than a year when I acquired a small Fowler toad, a fairly common eastern species, and placed it in the terrarium with the frog. Max simply ignored the toad, which was quite content to wriggle into a corner and half-cover itself with leaf mold and moss, so that little more than its head was visible.

This Fowler toad displays the normal mottled colors of its species.

Half covered by dry leaves and dark moss, the toad becomes almost indistinguishable by assuming the color of its background.

There could hardly have been a greater contrast than between Max and the little toad I named Bertha. He is as bold and unafraid as she was shy and fearful. The least movement near the terrarium caused her to duck her head and attempt to back even farther into her self-made burrow. While Max leaps to take a mealworm or a piece of fish from my fingers, sometimes even grabbing my fingertips in his eagerness, Bertha would retreat into the farthest, darkest corner, not even touching the worm or slug until I had withdrawn my hand, closed the lid of the terrarium, and waited quietly nearby without moving until the toad had calmed down again.

Although limited mostly to gray, brown, and olive hues, Bertha would change color every time I exposed her to a different background. Provided I allowed the animal sufficient time to settle down, and if the new environment was halfway acceptable by toad standards, the hues and patterns she displayed were almost invariably a close approximation of the general background color. Thus Bertha would turn dark brown in a container filled with leaf mold, blackish olive on sphagnum moss; and mottled brown and gray—her normal pattern— on a background of pebbles, earth, and dead leaves. In each case, the degree of resemblance to the substrate was remarkable; the animal was sometimes difficult to locate despite the small size of the terrarium, as its habit of remaining absolutely motionless enhanced the effectiveness of the camouflage coloring.

The results of these experiments were in striking contrast to others in which the toad was exposed to clearly unnatural environments. Whenever it was placed in a container that afforded it no place to hide

Disturbed and nervous, the toad keeps its pale "fright" coloration regardless of the color of the background to which it is exposed.

The same toad in a yellow-walled tank.

and bore no resemblance to a natural habitat, a paling of the toad's colors, which I learned is associated with fear and apprehension, almost immediately occurred. In this phase, the basic skin color lightened to a very pale gray, especially on the head, neck, and back; on this background, a number of light-rimmed brown spots and lines, normally not clearly defined, stood out in sharp relief.

Background or substrate colors could in no way influence this emotional color change. I tested this repeatedly by exposing the toad to several differently colored backgrounds over prolonged periods of time. Lining one container with black and another with white paper, I placed the toad alternately in each of them for several hours. In both cases the coloring paled in the way described above, and remained unchanged until I had transferred the animal once again to its terrarium, where within a short time it darkened and assumed its normal adaptive coloration. Tests with yellow and green containers yielded

93

the same results, which indicated clearly that background color was not a factor in these changes.

In order to find out whether such other factors as moisture, light, and temperature would affect the paling of the colors, I exposed the toad, in both the white and the black containers, to dampness, to very bright and very dim light, and to temperatures ranging from 50 to 85 degrees Fahrenheit. None of these variables proved to have any influence whatever on the pale "fear" coloration over periods up to six hours. In every case, however, transferral back to its familiar surroundings resulted in the animal's turning dark again within twenty minutes at the most.

From my admittedly limited observations of this toad, the apprehension or discomfort expressed by a paling of its color was the only instance of an emotional factor causing a color change in this particular animal. Judging from the example of Max and many other frogs I had kept over the years, I would have expected at least some color change during feeding time, which I never observed in Bertha. Possibly this was because the toad was much less active than the frog—at least during daytime—in the pursuit of its prey, usually preferring to wait, without moving far from its burrow, until the worm or slug had come close enough to be captured. However, even on those occasions when Bertha chose to pursue an insect, I never saw her change color. That was only one of the many behavioral differences between this toad and my green frog. Unfortunately, I was unable to continue my observations of Bertha, because one day, like the female frog before her, she decided to leave the enclosure on the lawn for the freedom and the dangers that lay in the garden and the woods beyond.

As one reviews the powers of color change in the various amphibians, it becomes clear that the arboreal species seem to be by far the most accomplished artists in this group, with the widest range of hues and the greatest mobility of expression. No ground-living frog can come even close to matching the feats performed by some of the *Hyla* species; the majority of the toads and true frogs appear to be more or less limited to a darkening or paling of their basic coloring. This, as a matter of fact, seems to be true of all terrestrial animals capable of color changes, reptiles and amphibians alike: those with powers of rapid change and a remarkable diversity of hues are for the most part arboreal.

On the whole, regardless of the degree of perfection, the ability of frogs and toads to change color adds another dimension to the faculties of self-expression in this group, as well as to our attempts at a better understanding of the humbler of our fellow creatures with which we share the earth.

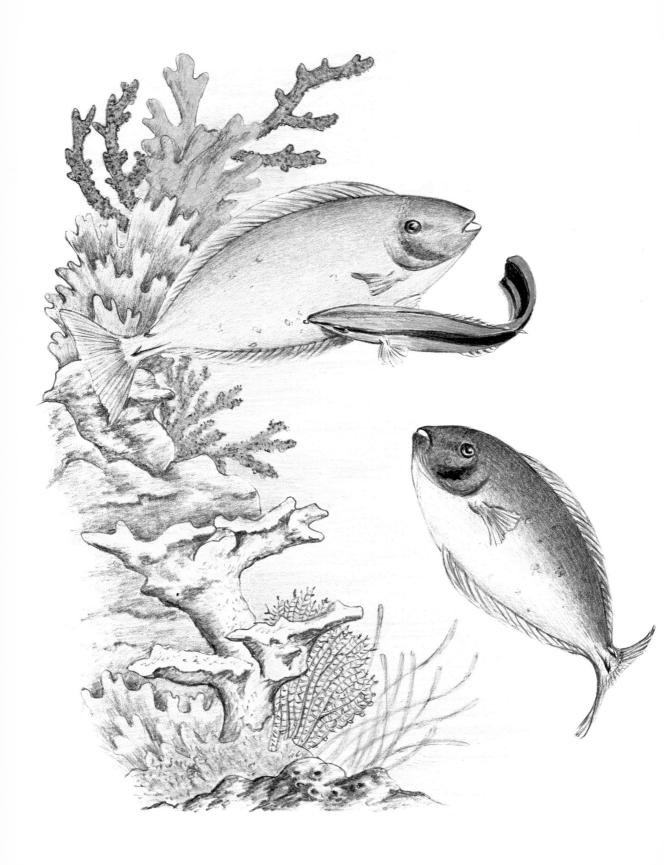

4. New Colors During Cleaning

IN THE clear waters of the reef, the unicorn fish stands almost motionless as a small, brightly colored wrasse approaches, circling the larger fish in a darting, dipping, tail-wriggling dance. Remaining stationary, almost as in a hypnotic trance, the unicorn fish spreads its fins. At the same time, its uniformly dark coloring begins to change until it has turned into a pale clear blue. On this background, a number of different-colored patches and spots stand out clearly. These patches, which represent a variety of bothersome parasites, are the target for attack by the wrasse. Methodically, the small black-and-blue striped fish works its way around the body of its larger "client," nibbling at the white patches until they disappear. The gills of the client fish, spread wide for easy access, receive special attention by the wrasse.

Finally, the operation is finished, and the unicorn fish moves, folding its fins. As it swims off, its color begins to lighten, changing back to its normal pattern. In the meantime the wrasse is already busy with its next client, for the species to which the little fish belongs is a member of a worldwide guild of distinctively liveried attendants that service countless "cleaning stations," especially in the warmer oceans around the globe.

The existence of cleaning stations that are regularly used is an important factor in maintaining the health of many species in reefs and other shallow waters. This was discovered only recently with the advent of diving devices that permit biologists to move about freely underwater and conduct prolonged observations of marine creatures in their natural environments and ecosystems. The mutually beneficial relationship called cleaning symbiosis between certain "cleaner" animals and their usually much larger clients has long been known to exist among land animals, and instances of this phenomenon were observed and recorded thousands of years ago. Cleaning symbiosis in the ocean, closely studied by a number of biologists in recent years, has been found to involve a much larger number of species, as well as a greater diversity of behavior patterns displayed by clients and cleaners alike, than in similar relationships observed on land. The majority of the cleaners in the water are fish, but certain species of shrimps and crabs also perform services of this type.

Among the most puzzling aspects of cleaning symbiosis is the fact that the cleaners are rarely, if ever, eaten by their clients despite the latters' larger size. The attendants seem to be safe even when they enter the mouth of the larger fish to clean between the teeth and gills, although the clients of the little cleaners include many voracious and

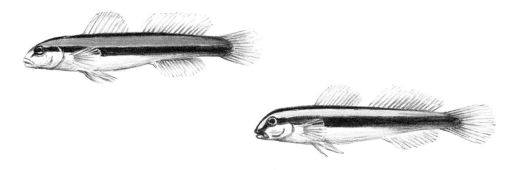

Two different but similar-looking cleaner fishes.

aggressive fish that normally would not hesitate to make a meal of any smaller fish that ventures conveniently close. Another enigma is the reason for the peculiar, almost trancelike state assumed by many client fish during the cleaning process, some of which have been observed in such odd positions as standing on their heads or lying on one side. In addition, the color changes which some fish undergo during the cleaning process still cannot be satisfactorily explained.

In the case of the discus fish, the color change may seem eminently logical, since the white fungus growths become very conspicuous against the dark background color, and hence are presumably easier to spot by the cleaner. But there is no solid evidence that the cleaner fish would be unable to find the fungus growths without the client's color change. Furthermore, many other fish undergo color changes during the cleaning operation that seem to be unrelated to the color of the parasites from which they seek deliverance at the cleaning stations. While being cleaned, some fish "blush," that is, they turn pink or reddish. Others turn pale. One reef fish from the Indian Ocean changes its almost black hue to a light, bright blue. What these changes signify, or why they occur, is still a matter of considerable speculation among naturalists, the best theory at the moment being that the changes are probably some kind of color code or signal meant for, and understood by, the respective cleaners. We must keep in mind that fish have excellent color vision—in most ways very similar to that of man— and that therefore color is important to them.

Although comparatively few fish capable of color changes have been studied critically, observation and research have shown that, of all animals capable of such changes, fish have the greatest diversity of mechanisms for achieving them.

Different types of color changes are common among fish. Slow, more or less semipermanent changes, for instance, which may take from several hours to days, or even weeks, permitting the animal to adapt itself to a particular surrounding, are comparatively rare among reptiles, but typical of many species of fish. Other changes, associated with courtship behavior, also may take many hours to complete. The majority, however, are more rapid and occur for a variety of reasons ranging from the expression of physical conditions, or of moods and emotions, to quick, "emergency" adaptation to a particular background.

Dispersal and aggregation of the pigment granules in the chromatophores are under the control of the nervous system, with special nerve

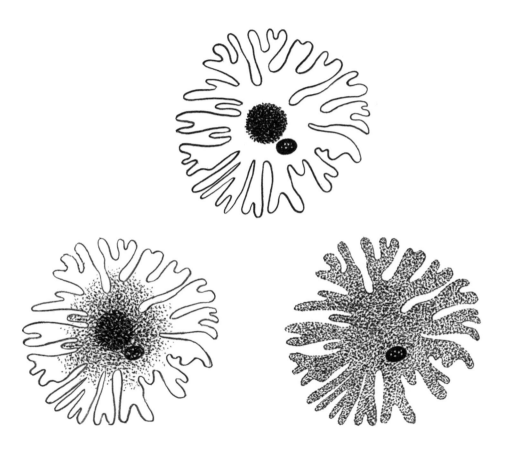

Melanophore of a fish showing the phases of pigment aggregation, partial and complete dispersal.

Micrograph of chromatophores demonstrating the intermediate steps between aggregation and dispersal.

fibers connecting the pigment cells to the nerve centers. In some cases, the nervous system stimulates the pituitary gland, causing it to release certain hormones into the bloodstream that disperse the melanin in the cells. Nervous stimuli sent along the sympathetic nervous system have the effect of aggregating the pigment. Most color changes in fish are controlled directly by the nervous system; they are more rapid than any of those produced by hormones. It appears that color changes affecting the color of the fish as a whole are caused more often by the somewhat slower hormonal action, while pattern formation is in most cases under direct nervous control, and therefore very rapid.

Fish may have black, white, red, yellow, and iridescent chromatophores, thus providing a wide and varied spectrum of hues. As in reptiles, the iridophores produce white as well as other structural colors through refraction and reflection of certain wave lengths; almost all blue and most green hues are produced in this way. In combination with red, yellow, and black pigments, a wide variety of hues becomes possible.

101

The ocellated blenny of European waters.

Many species that commonly live in coastal waters occur in different color phases adapted to particular environments. One prominent example is the demon stinger of Japan, *Inimicus japonicus*. As even its generic name proclaims, this member of the ray family is an unfriendly fellow that can deliver a potent sting. Found in various places along the shore, the demon stinger is blackish when living among dark lava rocks, but blood red, mottled with some brown, in waters where red algae occur.

An even more striking example of perfection in color adaptation to a particular environment is the band-shaped blenny of the California coast. Blennies are a worldwide group of fish that include such odd-looking forms as the ocellated blenny, or butterfly fish, of the Mediterranean and other European waters. These fish àre found mostly along rocky shores where different stone formations, as well as abundant aquatic plant life, may provide a wide range of different background and substrate color. The band-shaped blenny has managed to adapt itself admirably to these variations in the environment, for it occurs

102

along the California coast in no less than three radically different color phases—a bright blood red, an equally bright grass green, and a warm olive yellow Despite these apparently conspicuous hues, the fish blend nicely into their respective habitats: the red blenny, like the red demon stinger, lives among red algae; the green phase naturally occurs among green plants; while the yellow-colored blenny is found in places that have yellowish sand and rocks.

Given sufficient time and the right color background, each of these color phases can assume the coloration of either of the other two. However, if such a fish is introduced suddenly into an environment with a

The kelpfish, a relative of the blennies, turns reddish brown among brown algae.

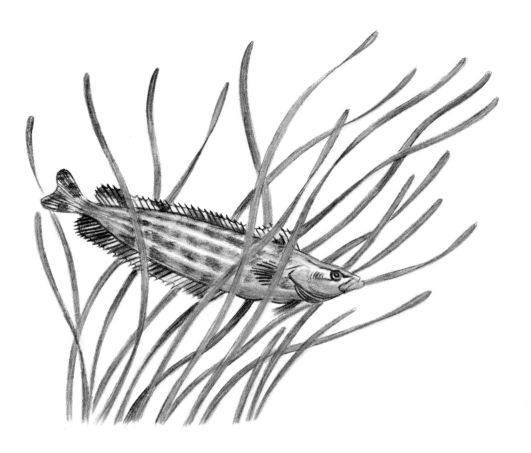

Among green algae, the kelpfish becomes and remains green. These adaptive changes are not rapid.

color scheme contrasting sharply with its own, it will dash for cover and try to hide, rather than expose itself, thereby proving that the fish is well aware that it has become conspicuous. This, by the way, is true of all fish that assume protective coloration. In some species, the changes take so long to complete that a "displaced" fish may seek an environment that fits its color rather than try to match an inappropriate background.

Other fish groups that specialize in slow, semipermanent color adaptations include the sticklebacks, pipefish, and wrasses. Sticklebacks are small fish that may live **in** both fresh and salt water. Some species

The odd-looking pipefish, a relative of the sea horse, among red algae.

While staying among green algae, the pipefish adapts to the color of its surroundings and assumes a shade of green as well as a position designed to hide
it from its enemies.

wander upriver from coastal regions; others prefer to stay in the ocean. All of them have a number of sharp, free-moving spines on their back that can be erected at will and that form an excellent defense against enemies, as well as a deadly weapon for the males in combat over territories. Sticklebacks often have semipermanent adaptive colorations that make members of the same species appear very different, but they are also capable of rapid color changes that reflect purely emotional responses, about which more will be said later. The aptly named long and slender pipefish are often found in light green, dark green, light brown, and brick red color phases, depending upon the color of the algae and the substrate occurring in their environment.

One intensive study of morphological color changes in fish in their natural habitat involved the small freshwater species *Gambusia affinis*, popularly known as the mosquito fish. This fish lives mostly in small lakes, ponds, and similar bodies of water, including some swamp and coastal waters that are quite brackish. Because it feeds voraciously on mosquito larvae, it has been used extensively in mosquito-abatement programs.

The mosquito fish shows great sensitivity to substratum coloration, having a light color phase when swimming over light-colored sand or gravel in shallow water, and a dark color phase when in deep water or over a muddy bottom. Such adaptive coloration affords this five-inch fish maximum protection in a habitat where its numerous enemies range from huge predatory water beetles to sharp-eyed kingfishers scanning the pond for food. The light coloring is necessary especially for pregnant females about to lay their eggs and for very young fish, both of which prefer shallow water because that is the best and safest place for the eggs to hatch and the hatchlings to remain in the early days of their life.

During weeks of observation, the mosquito fish were found to have a diurnal color rhythm that changes as the day advances, depending on the amount of light that penetrates the surface of the water. An-

The mosquito fish Gambusia *over light and dark substrates. In each case, the fish displays a shade that matches the substrate.*

other interesting sidelight was that fish forced to remain for prolonged periods over dark substrata become more susceptible to infectious diseases, so that we can assume that the color rhythm is important for maintaining the health of this species.

In this study the mosquito fish were observed under controlled conditions in carefully selected ponds but in their natural habitat. Most research of this type involves observations of fish in their natural environment without a chance of controlling any of the prevailing conditions, or experiments with fish kept in tanks—both yielding valuable if partial results. One such study in the ocean showed that many fish change their coloring according to the depth in which they swim, appearing light blue or bluish gray when near the surface, and yellow or blackish brown—depending on the substrate—when they stay close to the bottom. Two such distinct color phases of one species were for a long time thought to be two distinct species, because one was slate blue, the other, yellow. Later, the error was discovered by observers who found that the yellow color was assumed by fish that stayed near the sandy or gravelly bottom of the coast, the slate blue hue by those that were swimming freely. Such adaptive color changes are frequent. The big surgeonfish, for example, is almost invariably black when near the bottom, but pale blue gray when found swimming twenty-five feet higher up.

A skate in its normal brown coloration.

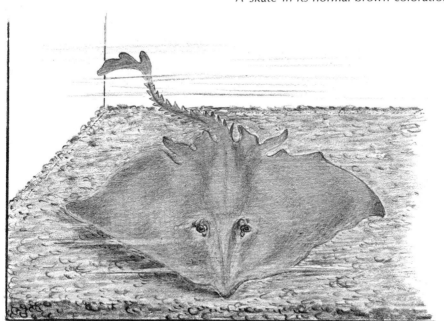

Other fish may employ yet another device to lessen the chance of becoming too conspicuous: they display a pattern of longitudinal stripes when moving, but one of transverse bars when resting near the ocean floor, thus creating an interesting optical illusion that helps protect them.

Experiments with fish in specially prepared tanks have yielded equally interesting results. In one such study, several skates were placed in a tank whose walls and bottom had been painted white, and others in a similar tank painted black. Skates, which belong to the ray family and share with the flatfish a shape that looks as though a steamroller had passed over them, normally have a color ranging from gray or grayish brown to slate blue. In the prepared tanks, the skates gradually changed, shade by shade, in an effort to adapt to the starkly extreme—

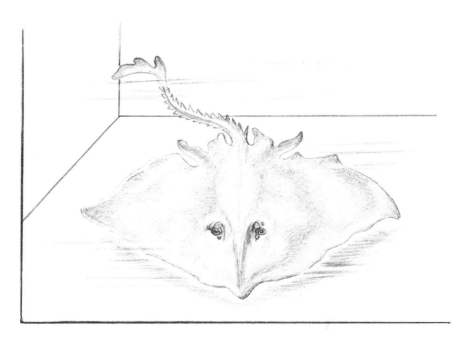

A normally brown skate kept in a white-walled tank turns pale.

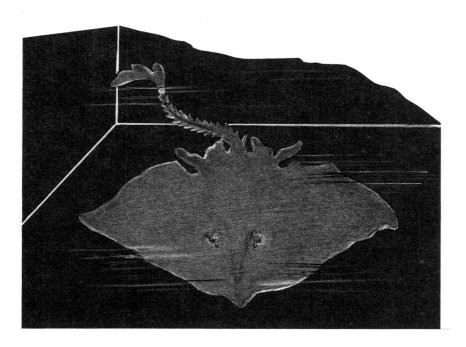

The same skate after a day in a black-walled tank.

and highly artificial—colors of their respective tanks. Some eighteen to twenty hours later, the skates in the white tank had turned a very light, creamy brown hue, those in the black tank, an almost blackish brown. These dark skates were then placed in the white tank, with the result that twenty-one hours later, the fish had become pinkish white.

The fact that light has a great influence on these color changes is demonstrated by the nighttime coloration which many fish display, and which in most cases differs considerably from their typical daytime coloring. Usually, the fish become pale and colorless—except for structural colors that remain unaffected because of their special nature. A good example is the cardinal fish, which during daytime has a rosy hue, washed with a glittering, bright iridescent blue. A sudden light, flashed at a cardinal fish during a nighttime dive, revealed that it had lost every trace of red and was a very pale light gray or beige, though still washed with the same iridescent blue. Exposing the fish to light for a prolonged period caused them to "blush" while the red pigment granules in the cells dispersed until the fish had attained their normal rosy daytime hue.

Slow, gradual adaptive color changes, such as those observed in the skates and many other species, would be of no help at all when rapid color changes become necessary. The ability to change quickly in order to merge with their surroundings is found in a great number of different types of fish, and serves them well especially in emergency situations where gradual, semipermanent changes would be quite useless.

One species capable of rapid color adaptation is the shark sucker, belonging to the small group of remoras. Although these interesting

A cardinal fish displays its daytime coloration.

The cardinal fish clad in its pale nighttime colors. Exposure to artificial light brought about a change to the rosy daytime hues within a few minutes.

fish do not suck blood from the sharks, as the name implies, they frequently attach themselves to these large predators and "bum a ride," a practice that has the double advantage of conserving energy and of providing excellent protection against most enemies, few of which would be bold enough to venture close to a shark. There is some evidence that the remora pays for the ride, however, for external parasites known to infest sharks have been found in the stomachs of captured shark suckers. This seems to indicate that the remoras reimburse their living transports by acting as cleaners.

The special feature that permits the remora to cling to the bodies of larger fish—or even to such surfaces as ships' hulls—is a suctorial disk located at the top of its head. A flat, transversely furrowed oval plate, this disk is actually a greatly modified dorsal fin. The membranes that make up the furrows create small vacuum pockets when pressed against

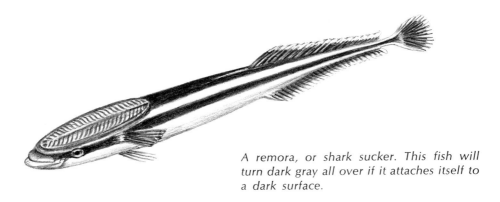

A remora, or shark sucker. This fish will turn dark gray all over if it attaches itself to a dark surface.

any reasonably flat and smooth surface, such as the skin of a shark or the hull of a ship. While the species known as shark sucker seems to prefer these large killers, other species often select different transports, a favorite being the huge sea turtles.

The shark sucker is about two feet long, whitish gray below, slate or blue gray above, with a white line running the length of the body on either side—a common coloration for any fish found in deep waters. If the remora attaches itself to the light underside of a shark, it will maintain its normal coloration. Not so if it happens to fasten its disk to the dark gray blue side of the larger fish: now the passenger changes color to match the vehicle, because its light-colored underside would be conspicuous. Within minutes, the remora has turned dark gray below so that it is dark all over, its color blending perfectly with that of the shark.

Probably the most versatile of all quick-change artists are found among the flatfish, which include the soles, halibuts, and flounders. Tests have shown that some flounders not only do their very best to match the *color* of the surface upon which they are placed, but also attempt to duplicate the *pattern* as well, and usually very successfully. A favorite test was to place such fish on three different backgrounds: one was a solid sand color, the second was mottled and pebbled in different colors, the third, a black-and-white checkerboard surface. Although it was not surprising that the flounder could adapt readily to the sand-colored background, matching it to perfection, the degree to which it simulated its pebbled, multicolored background was astonishing indeed. Most incredible, however, was the attempt by the fish to match the checkerboard substrate: although it could not, of course, form actual squares on its skin, the fish produced a very remarkable black-and-white approximation of its geometric and quite unnatural background.

All color changes in fish that are adaptations to the substrate are brought about through the eyes, and to a lesser extent through certain

Excitement transforms this European relative of the killifish, changing its light color to a dark mottled pattern.

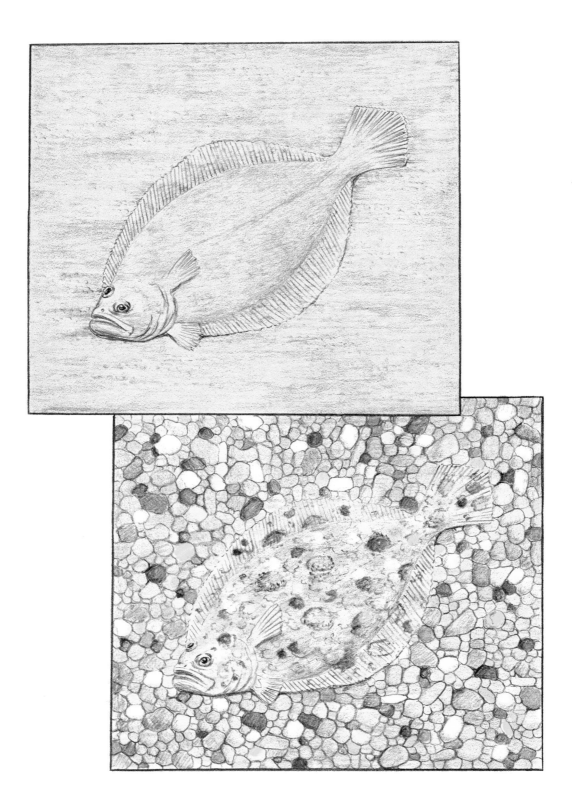

light perceptors in the skin. That the eye is the most important organ for achieving these changes was proved by placing a flounder on a background consisting of two sharply contrasting colors: the fish assumed the color of that portion of the field which it was able to see.

It is a different matter entirely with color changes that take place for reasons other than adaptation, especially those changes expressing moods and emotions and tending to make the fish more, rather than less, conspicuous. Instead of being reactions to the colors the fish perceives around it, such changes are responses to what it feels— fear, excitement, anger, or triumph—and consequently have to be quick. A fish

A flounder on three radically different substrates: in each case, the fish tries to adapt to both the color and pattern of the background.

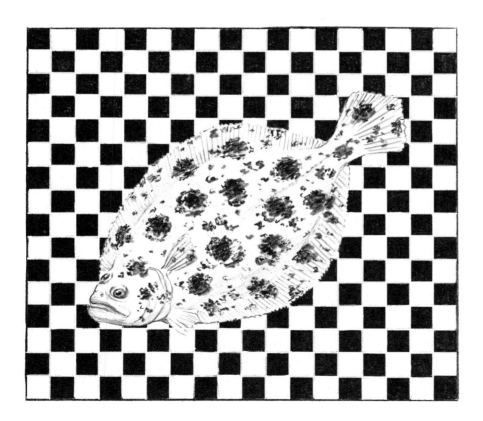

that wants to express a temporary and fleeting mood can hardly wait hours—or even half an hour—for the appropriate color change to occur. Accordingly, the majority of emotional color changes in fish take from just a few seconds to a few minutes at the most. One of the truly accomplished quick-change artists is the osbeck, a fish found in the waters off Bermuda. Observers clocked this fish and found that it needs only between four and eight seconds to change from a predominantly red pattern to almost white. Others are not quite that fast, but very rapid kaleidoscopic changes in frightened or excited fish are common, especially among the inhabitants of coral reefs, whose powers of rapid color change far exceed those of any lizard, including the chameleon.

An outstanding example is the large Nassau grouper, which may

A Nassau grouper displaying a typically mottled pattern.

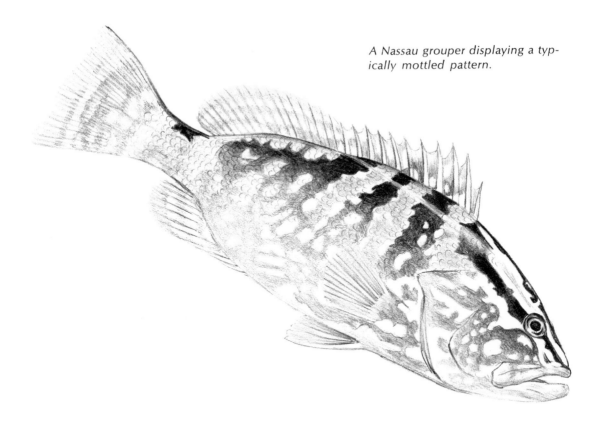

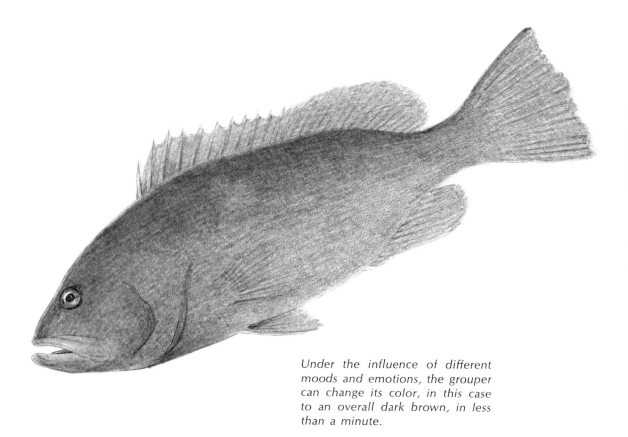

Under the influence of different moods and emotions, the grouper can change its color, in this case to an overall dark brown, in less than a minute.

appear in at least six different liveries, each with its distinctive colors and patterns, putting them on and off in a matter of just a few moments. When startled or frightened, this fish assumes a bold pattern of dark stripes and patches on a light-colored background. Its color repertory ranges all the way from creamy white to a blackish brown, with innumerable hues and shades in between. One marine biologist observed that, whenever a closely related red grouper came near, the Nassau grouper displayed a color phase not seen at other times, almost reversing its normal coloring to the point where the distinctive dark stripe through the eye became almost creamy white. So far, no explanation of this particular change has been offered by scientists.

119

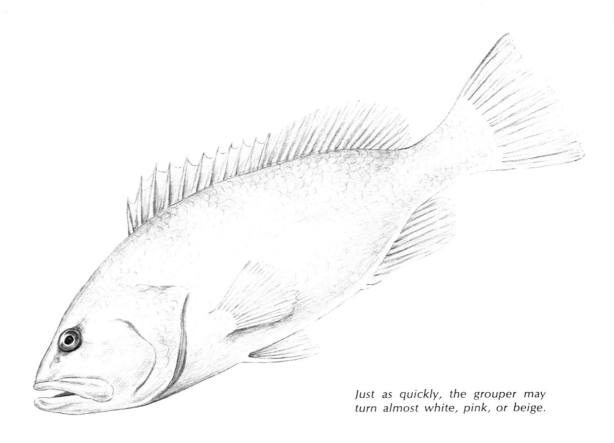

Just as quickly, the grouper may turn almost white, pink, or beige.

Color changes in fish associated with mating rituals during the breeding season have been studied perhaps more closely than any other, mainly because they occur in many fish that can easily be kept and bred in captivity, and may therefore be observed at leisure and under favorable conditions. Prominent among these are the sticklebacks, mentioned earlier in this chapter. One of the most common European species is the three-spined stickleback, which has three hard, sharp spines on the back and two small ones on the underside. Thus armed, these fish can defend themselves against even large predators, and accordingly are bold and aggressive animals despite their small size of just three inches. Reputedly, even such rapacious fish as pikes respect the sticklebacks' weapons and leave them alone.

Common in many rivers, ponds, and coastal areas of western and northern Europe, the stickleback is often kept in captivity because of its attractive coloring, vivacious behavior, and habit of building beautifully constructed nests for its eggs. The normal coloring for this species ranges from greenish brown to blue black above, and silvery on the sides and below, with the exception of the anterior portion of the underside, which in the males is a beautiful rosy red. These colors are subject to great variations depending on the environment in which the fish are found, and to great changes during the mating season, when fighting among the males and courtship behavior bring about rapidly shifting hues.

I started to keep sticklebacks when I was a child, and enjoyed observing them for hours, watching with wonder as the colors gleamed and shifted, intensified and paled during combat between the males. At that time, I did not yet know enough to realize that the vicious fighting among my fish was caused partly by the limited space in which

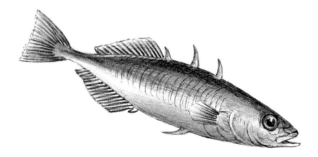

Two male three-spined sticklebacks after a territorial battle: the victorious male displays brilliant colors, the loser turns pale.

they were confined, making their individual territories touch or overlap and thus provoking fierce territorial fights. I only grieved when one of the males, badly hurt in the fighting, rapidly paled to the grayish color of defeat, and died the next day despite my attempts to revive him. In later years, I learned more about these fish and always took care to allow them enough room. With fascination I watched as the males dug holes in the sand and then began to build their nests. I was not successful, however, in raising a brood in captivity, probably because the environment in the tank lacked some ingredient necessary for successful breeding.

Normally, the mating sequence starts with the male staking out and defending a territory. In their natural environment, the males fight less than in captivity because the territories are spaced farther apart. However, any male that infringes upon the boundaries of another is in for

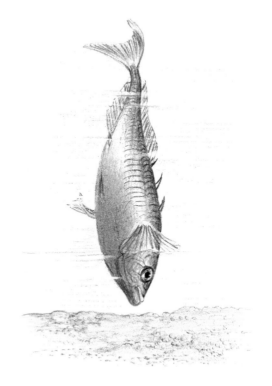

Threat display by a male three-spined stickleback. The bright red underside is turned toward the adversary.

a battle that may well end with the death of one of them. As they fight, they also shift color with incredible rapidity. The victor is a sight to behold: the back turns a deep green, set off beautifully by the now blood red throat and breast and the silvery white of the sides and the belly. The loser, regardless of whether he is hurt or only scared away, turns pale. Very often during these battles, threat displays are employed. A male stands on its head while making movements as though it wants to pick up something from the bottom. At the same time, the body is turned toward its opponent in a way that exposes as much as possible of its red underparts.

As soon as the territory is successfully established, the male begins to prepare for the rearing of the young, for among sticklebacks this is the exclusive duty of the male. The female does nothing at all for her progeny except lay the eggs.

Nestbuilding begins with the excavation of a suitable hollow in the sand. Standing on its head, the male digs with a wriggling motion until it has made a fairly deep depression, and then starts to construct the actual nest—a long and arduous task. The gathering of suitable material may take only a few hours if it is plentiful. All kinds of plant fibers are used, including rootlets of aquatic plants. The stickleback tests all pieces by seeing whether they float or sink to the bottom: the former are discarded, the latter used for construction. After arranging each layer, the fish glues the strands together with a kidney secretion. Meticulously woven, the nest may take days to construct. Finally, it is finished: a roughly egg-shaped affair about the size of a small fist, with an entrance on one side, but completely closed at the top.

Now the male seeks a female and induces her to enter the nest, where she lays her eggs, then pushes her way through the other side, and swims off. With the nest open at both ends, water can flow freely over the eggs, a process encouraged by the male, who stands guard at the entrance, busily fanning water through the opening with his fins. In this way, the eggs are assured of a constantly replenished oxygen

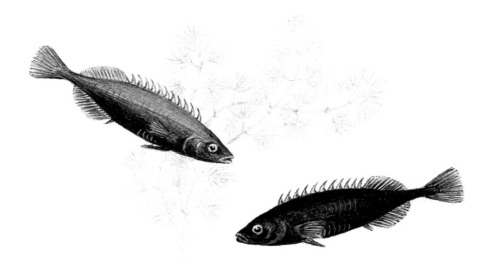

A male eleven-spined stickleback in its courtship colors (top) and in its all-black battle dress.

supply. Still dressed in his brilliant mating colors, the male keeps a sharp eye out for possible enemies and fiercely attacks anything that ventures too close. When the eggs finally hatch, the anxious father continues his vigil, keeping the young close together in the by now somewhat damaged nest. Predators with a taste for young sticklebacks at this early age include their fond but voracious mother, who is driven off by the male with the same fury as any other enemy lusting for a tender young fish. Only after the young have become accomplished swimmers and are able to fend for themselves does the father's solicitous concern for them begin to wane; a little later, the group breaks up altogether, with the young seeking their own living and their father resuming his normal life style, once again dressed in his everyday colors.

Other sticklebacks display similar behavior, and in all species the color changes that accompany the various activities of the breeding season are marked. The tiny eleven-spined stickleback, only a little over two inches long, is normally greenish above and silvery below. When fighting with another male, however, it turns a deep velvety black all over. Black is also the courtship color of the species, but it is easy to tell whether a male is going courting or fighting, for the courting male turns black only below, retaining its greenish hue above.

Black seems to be the preferred color of a number of fish in their courting activities—the smart thing to wear—especially for many of those species that engage either in nest building or in other types of special care for their young. One of these is the river bullhead of England, which builds its nest in a crevice and later guards the eggs, very much in the manner of the sticklebacks. The most conspicuous feature of this fish is its large, wide head, which is the part that turns completely black during fighting, as well as during courtship activities. The rest of the body retains the normal, much lighter coloring.

One of the most extensive studies on color changes in fish during the breeding season involved a member of the cichlid family. Cichlids, which include many favorites of aquarium hobbyists, are a group of tropical freshwater fish related to the basses and perches, and many members are noted for their bright coloring, as well as for their powers of rapid color change. The species of the study, *Tilapia mossambica*, a cichlid from the east coast of Africa, now is widely known as a valuable food fish in southeast Asia, where it was introduced into ponds and flooded rice fields and has since thrived.

Tilapia mossambica *in neutral coloration.*

The ability of *Tilapia* to change color rapidly was one reason why this fish was selected for the study. It was well known that when these cichlids are tranquil—swimming, resting, or feeding—they generally assume the color of the substrate as long as they remain unmolested. In a green tank *Tilapia* will turn green; in a white-walled enclosure, cream colored; and black in a black-walled tank. These adaptive changes occur within two hours. However, if one of these fish is startled or chased, it rapidly changes to a barred or striped pattern, and if the disturbance continues, it develops a hatched design of crisscrossing bars and stripes. These changes take only a couple of minutes—a very dark male when chased blanches in less than two minutes to the hatched "disturbance" pattern.

During the breeding season, the color changes of *Tilapia* fish free from outside disturbance have an entirely different rhythm, beginning with the fight for territories, which takes place mainly among males

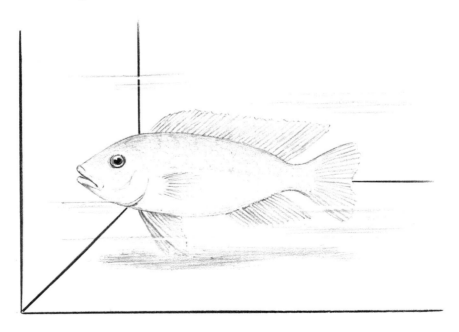

Tilapia *adapts to the color of its surroundings by turning pale in a white-walled enclosure . . .*

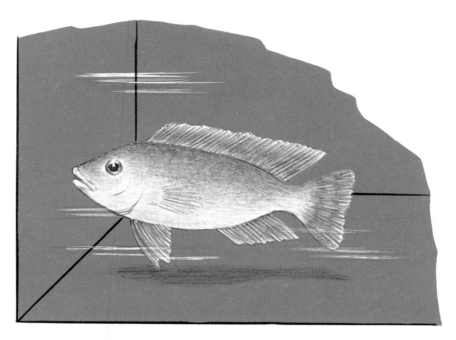

. . . and green in a green-walled tank.

but which also may involve females once a territory has been established. Fighting males turn dark gray—but not black—while they stand their ground. The moment one of them turns to flee, however, it turns pale, thereby signaling that it is conceding the fight to the other male.

Pregnant females assume a dark silvery color, which changes very little, but the courting male goes through a series of changes that basically consist of a darkening of its overall color until it has turned a very dark gray. The next phase is distinctive: a velvety black marked with blue stripes. This is the next to the last in the courtship color phases, although it appears that many males mate in this garb and do not change to the ultimate breeding pattern. Others, however, go beyond the blue-banded black stage to display an overall rich velvety black, set off by a brown head, white jaw, and red-tipped dorsal and caudal fins. Because these colors are caused by sex hormones, the changes are relatively slow: it takes about twenty-four hours to attain the all-black phase.

5. The Blushing Octopus

To THE visitors to the famous aquarium in the Berlin zoo, the large, beautifully appointed tank, with its cleverly hidden, dim lighting that reveals well-arranged groupings of rocks and stones and a variety of algae, presents an authentic reproduction of a small portion of the underwater world along a rocky coast in the Mediterranean. The animals in the tank, ranging from sea anemones to fish, seem very much at home in this environment, and so does the small octopus, a familiar European cephalopod, which lies draped along a rocky ledge, its tentacles trailing on the sand below. The animal's coloring is a pale grayish tan, mottled with some darker blotches. Its almost human looking eyes are watchful, apparently missing little of what happens inside the tank, and at the same time alert to movements outside. Suddenly, something excites or disturbs the octopus: in one smooth sensuous motion, it gathers its tentacles and half-slides, half-scrambles along the ledge toward the far end of the tank, where it now comes to rest in a crevice between two large rocks.

While in motion, the cephalopod began to turn a dull rose color, which spread rapidly over its entire body in undulating waves, replacing within seconds the grayish hue the animal displayed earlier. At the same time, the skin texture changed, becoming alternately rough and warty, and then smooth again in a rapid play of muscle action. Then, after a few minutes of repose, its neutral light gray coloration reappeared.

It was clear to any observer that the octopus's sudden color change was in no way related to, or influenced by, the color of its environment. Obviously, the animal was expressing some mood or emotion. Fear, discomfort, excitement—it could have been any one of these, for the emotional "color-change code" of this most advanced of marine invertebrates is still incompletely understood.

Color changes are frequently found in other invertebrates besides cephalopods. Some echinoderms and crustaceans may be limited to a mere darkening or lightening of the basic coloring, while others, and especially certain cephalopods, are capable of the most sophisticated types of color change found in any species, whether vertebrate or invertebrate, terrestrial or aquatic.

A sea urchin appears dark brown in the daytime.

At night, the sea urchin changes to a pale blue color.

Simple color changes triggered by variations in light are typical of most sea urchins, the peculiar echinoderms many of which resemble hedgehogs because they are covered with movable spines. These sea urchins darken when illuminated, and become paler as the light fades. A definite daily rhythm of color change has been observed in some species. Very pale at night, they begin to darken as the light filters through the water after sunrise, until their normal daylight coloration is attained.

The intermediate phases between light and dark in certain echinoderms are distinguished by the transitory appearance of brilliant blue hues—structural colors caused by dispersion and reflection of light as the dark pigment recedes. As most of the component colors of white light are absorbed by the melanin background, only the shortest wave lengths are reflected, producing a pure, nonpigment blue similar to that of the sky on a clear summer day.

Among crustaceans a wide variety of groups including the shrimps and crabs have members capable of rapid color change. In some more primitive species, such as the isopods, these changes, like those found in sea urchins, consist mainly of a darkening or lightening of the basic hue, depending upon the amount of light exposure.

Some higher crustaceans, on the other hand, are extremely versatile and capable of spectacular color changes. These species have been objects of interest and study for generations of biologists. Although painstaking research has revealed much about the biochemical and physiological nature of these changes and identified some of the stimuli that trigger them, many other aspects still remain as enigmatic as ever. There is, for example, the puzzling fact that one crustacean may change color exclusively for environmental adaptation, while a closely related species living in a similar environment displays different hues solely to express moods and emotions, and never for camouflage.

Many kinds of shrimps and prawns are typical of those crustaceans that adjust to their surroundings by displaying the appropriate hue and pattern; their eyes perceive the colors of the environment and trigger the adaptive change. The classic case is the small prawn *Hippolyte,*

At night, Hippolyte *assumes a pale, transparent blue coloring.*

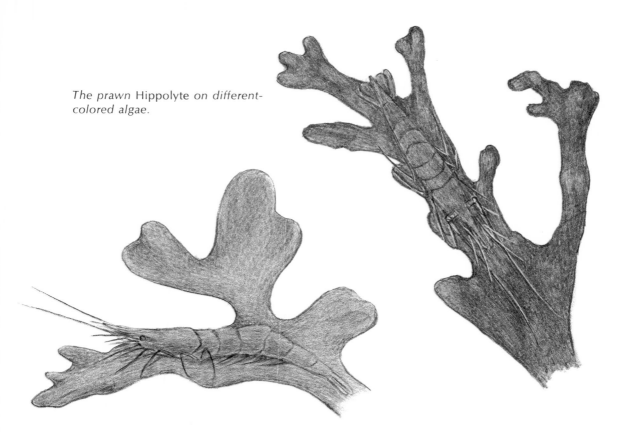

The prawn Hippolyte *on different-colored algae.*

famous among naturalists because it has been the object of intensive study in several countries in the past fifty years, making it one of the most thoroughly investigated instances of rapid color change.

Like most animals capable of altering their appearance in this manner, *Hippolyte* has a definite color rhythm influenced by light. At night, the pigment in all the chromatophores aggregates, and the prawn turns a transparent nonpigment blue. During daytime, however, *Hippolyte* may display any one, or several simultaneously, of a wide range of colors and patterns, most of which match its environment to a remarkable degree. Among green seaweed, for instance, *Hippolyte* turns exactly the same shade of green; the prawn will be olive colored if the seaweed is olive colored, and brown if brown algae happen to predominate in its habitat. Among coraline algae, on the other hand, it assumes the exact violet shade of these plants. Because of its extensive repertory, it may

135

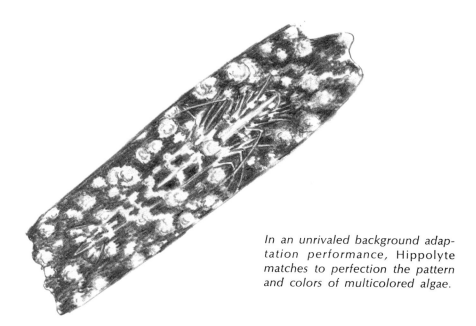

In an unrivaled background adaptation performance, Hippolyte matches to perfection the pattern and colors of multicolored algae.

also turn yellow, orange, or red if any of these colors are required to make the prawn blend into its surroundings. The most fantastic changes occur whenever *Hippolyte* decides to settle on a growth of peculiar, multicolored algae with a pattern of hues that range from pink to violet. The prawn is capable of matching not only the various colors, but also the asymmetrical pattern of these growths so perfectly that the animal is all but invisible. In a game of Find the Hidden Hippolyte, even a sharp-eyed observer could not spot the prawn without considerable searching.

The advantage of such perfect disguises by color change over those in which animals are limited, by body pattern and color, to mimicking a single object—a leaf or blossom, for instance—is the freedom of choosing a variety of different habitats. This freedom is an important aid in the struggle for existence, which is even more relentless in the ocean than on the land.

The speed of the color changes in *Hippolyte* depends entirely upon the individual prawn's age. The younger the prawn is, the faster it can change its garb. As mentioned earlier, many fish also can change color rapidly while young, but lose the ability to a certain extent as they grow older. A young *Hippolyte* may need only ten minutes for a complete change, while at later stages it requires from ten to twenty-four hours. Mature animals have an added difficulty: they must wait for one of their periodical molts before changing color. It is therefore quite understandable that older prawns would much rather move than change—meaning that they prefer to look for an environment that fits their color, rather than to wait for the slow adaptive process to take effect.

An interesting contrast to the purely adaptive changes in shrimps and prawns is provided by those of a different type of crustacean, the fiddler crab. This small crab belongs to the burrowing crabs, and is fairly common along the Atlantic coast. Despite its small size, the fiddler crab manages to dig holes in the sand that may be up to three feet long.

The male fiddler crab has one tremendously enlarged claw, which it holds in a position reminiscent of someone playing the violin; hence the

A fiddler crab displaying its normal coloring.

An excited or frightened crab may turn bright red . . .

animal's name. This menacing weapon is used mainly in battles between rival males during the mating season, but rarely to inflict serious injury. Occasionally the male may barricade its burrow by covering the entrance with his huge claw.

Normally the fiddler crab has a light brownish coloring, but that hue is subject to rapid and radical changes. With the animal possessed of a quick temper and the means for reflecting every mood, its coloring may lighten or darken, assuming red, purplish, or blackish hues as its owner runs the gamut of emotions from sexual excitement to anger. The color of the environment, so important a factor in many other crustacean color changes, has no influence at all on those of the fiddler crab, nor is the animal restrained by the fact that its emotional color outbursts may make it conspicuous and easy to detect. Other burrowing crabs, on the other hand, seem to prefer adaptive color changes—usually a simple darkening or lightening of their normal coloration—and do not indulge in the fiddler crab's type of emotionalism.

138

The chromatophores involved in crustacean color changes, though very much like those of reptiles and fish capable of such changes, are frequently quite complex, in some cases containing four different pigments in a single cell. Because each of these pigments may be activated by a different stimulus, a wide variety of hues as well as great mobility of change is possible.

Until the early 1900's it was believed that the chromatophores of crustaceans were under direct nervous control. More recent research led to the discovery of two glands, the so-called sinus gland and X-organ, both located in the eyestalks. Apparently these glands are hormone-gathering, rather than hormone-producing, organs, and are directly responsible for activating the pigment granules in the chromatophores. The process of color change in a shrimp, for example, starts with the eye of the animal responding to the color of its surroundings. This stimulates the release of hormones from the sinus gland and the X-organ into the blood stream, which in turn stimulate the chromatophores either to aggregate or to disperse their pigments. To complicate matters, it has been established that color changes influenced by

. . . or dark within seconds.

the upper half of the crustacean's compound eye differ from those triggered by the lower half.

Extracts from the glands were used in a variety of tests, all of which proved conclusively that the aggregation and dispersion of pigment in crustacean chromatophores are activated by endocrine secretions. For example, extract from the eyestalk glands of a shrimp, injected into another shrimp, caused a paling of the latter's color. The same hormone injected into a fiddler crab, however, activated the melanin-dispersing hormones in the crab's eyestalks and made it turn dark. When blood taken from a dark shrimp kept on a black background was injected into a light-colored shrimp on a white background, the light shrimp darkened, although the background color remained white. Blood from a shrimp that had turned yellowish on a yellow background, when injected into a light shrimp kept on white, caused it to turn yellow. However, attempts to reverse this procedure failed: tests in which blood

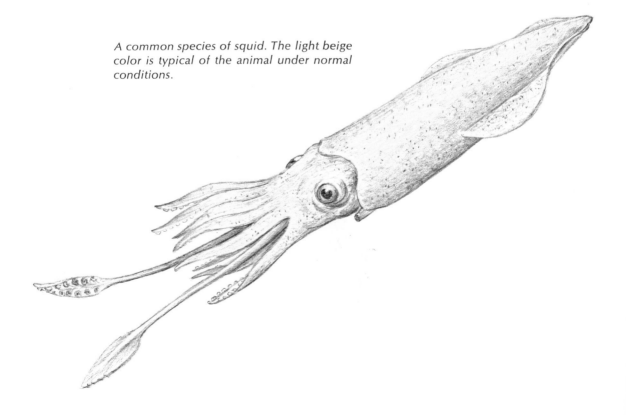

A common species of squid. The light beige color is typical of the animal under normal conditions.

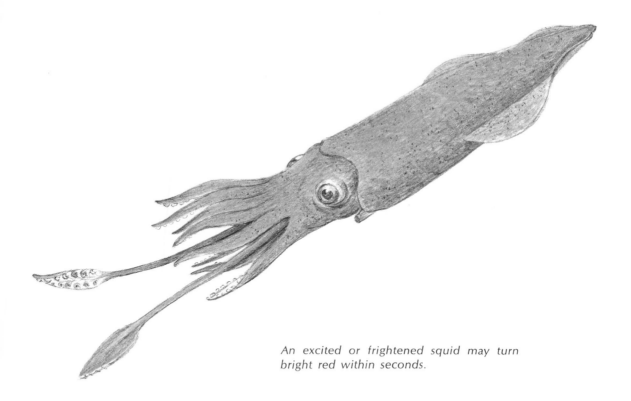

An excited or frightened squid may turn bright red within seconds.

from a light shrimp was injected into dark shrimp were negative, resulting in no color change at all.

Unsurpassed in both the range of their color repertory and the rate of speed with which they can change are the cephalopods—the octopuses, squids, and cuttlefish. The rainbowlike displays of rapidly shifting hues that members of this group produce are considered by some observers the most marvelous of all such changes found in the animal world.

Many people have difficulty in distinguishing between the three major groups of cephalopods. All these most highly organized marine invertebrates are mollusks in which the foot has been modified into a ring of tentacles, or "arms," that bear cuplike suction disks. Octopuses have eight arms; squids and cuttlefish, ten. The main distinction be-

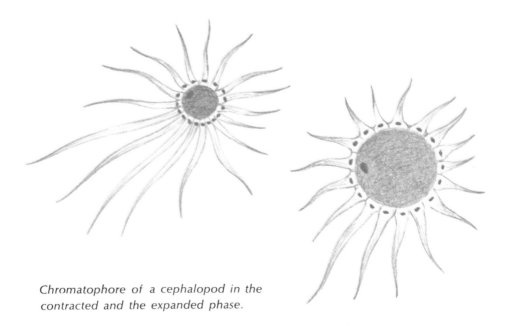

Chromatophore of a cephalopod in the
contracted and the expanded phase.

tween the ten-armed—or ten-legged—groups is the hard, calcified internal shell of the cuttlefish, which in the slenderer squids is replaced by
a soft, horny plate called the pen. All of these animals are able to move
by jet propulsion; squids especially swim by ejecting a stream of water
that propels them backward, a mode of locomotion that has decided ad-
vantages when the animal needs to escape from an enemy.

Most cephalopods have a very special protective device, consisting of
a sac filled with an inky fluid which they emit when feeling threatened
or pursued. Disappearing in clouds of inky water after discharging their
secretions, these inventors of the original smoke screen can often make
their getaway while their enemy may be literally groping in the dark.

The secretions of various cuttlefish yield a rich-brown coloring matter used especially in watercolor painting. This shade of brown, known
as sepia, was named after the cuttlefish genus *Sepia*.

With the exception of the mythical sea snake, few ocean creatures
have figured so prominently in adventure stories, told by sailors around

the world, as huge and ferocious squids and octopuses that attack ships and men. Giant cephalopods do exist, but they are rare, found almost exclusively in deep water. We know from scars on captured sperm whales and from other evidence that battles between these huge whales and giant squids occur from time to time in the ocean depths. There are also a few reliable eye-witness reports of giant octopuses, but none has ever been caught alive, and no captured specimen has been as big as some sailors' yarns make them out to be. One of the more interesting sightings of a giant squid was reported by the captain of a French dispatch boat in the year 1861, when he and his men sighted the huge cephalopod in the ocean between the islands of Madeira and Tenerife. The captain estimated the squid's body length—not including the tentacles—to be about eighteen feet, and its eyes were immense. When the crew began to attack the animal, it turned brick red. All attempts at capturing the monster were in vain; the captain did not dare to lower a boat, fearing that the squid would capsize it and kill the crew. Finally, after a three-hour battle, all the men had to show for their efforts was a small piece of the squid's mantle.

Despite such dramatic tales, even a five- or six-foot cephalopod is exceptionally large, as the majority of these animals are less than two feet long. They are generally shy and harmless, although some have venomous salivary glands and can inflict a painful, and occasionally even a fatal, wound. Usually, however, they try to avoid encounters with man in the ocean, and restrict their attacks to the small crabs and shellfish on which they feed. Many cephalopods are prized as food; especially on the coast of the Mediterranean, squid and octopus, along with other *frutti di mare*, are regularly sold in open-air fish markets.

When they are tranquil and resting, squids and similar cephalopods may adapt their coloring to a certain extent to that of their surroundings, most often displaying the relatively neutral colors of the sand, pebbles, and rocks among which they like to stay. A mottled sand coloring, a disruptive pattern of light and dark blotches, and a yellowish

143

pink hue mottled with white spots—all three are favorite resting colors of a familiar European cuttlefish. But let the animal be disturbed, or become angry or excited, and other diverse hues and patterns immediately replace these neutral colorations. The fright pattern of this species consists of two huge blackish brown spots like a pair of monstrous eyes that appear on the animal's back, while the rest of the body displays a variety of mottled, shifting, occasionally iridescent hues accented by the altered skin texture, which changes from smooth to rough and warty. Angry or excited, the same cuttlefish takes on reddish or dark brown hues with coppery, iridescent overtones.

Common European cuttlefish in three different color phases. At left, top, the breeding pattern of the male. Below, resting, or neutral, coloring. On the right, the animal displays the fright pattern.

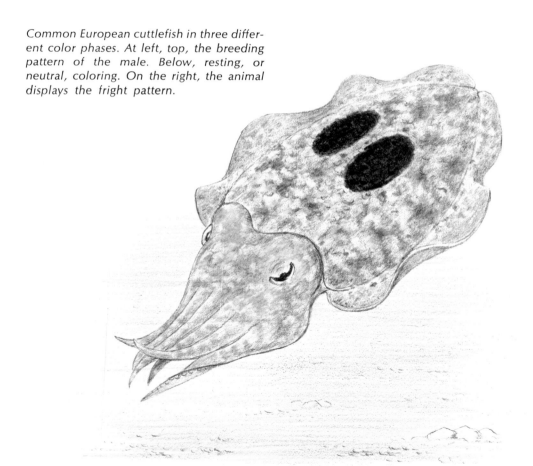

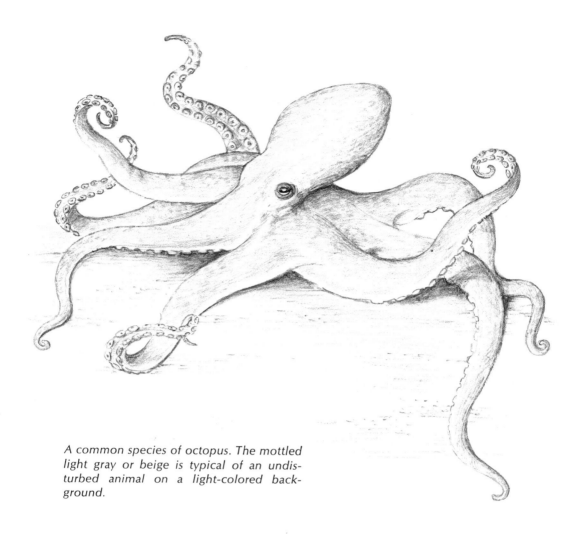

A common species of octopus. The mottled light gray or beige is typical of an undisturbed animal on a light-colored background.

During the mating season, this cuttlefish appears in a garb that differs from any of the others, a distinctive pattern of black-and-white zebra stripes worn only at courting time.

Many octopuses change primarily to hues in the yellow-orange-red-brown range when expressing different moods. Thus an octopus displaying a grayish white color one moment may "blush" a reddish tint the next, as described at the beginning of this chapter. Especially during the excitement of hunting and capturing prey, octopuses tend to change color rapidly, shifting from one hue to another until their hunger is satisfied and they finally return to their neutral resting coloration.

Even a century ago, the fact that cephalopod chromatophores are unique, differing from any found in other animals capable of color changes, was recognized. In the typical chromatophore, the pigment is either aggregated in the center, or dispersed throughout the cell upon stimulation, causing the animal's skin to lighten or darken, respectively, without affecting either the shape or the size of the cell. Cephalopod chromatophores, on the other hand, actually change their shape *and* their size: upon receiving the nervous stimulation, the cell is either contracted or expanded by numerous radially arranged muscle fibers. In the cell's contracted, starlike form, the pigment is squeezed into a tight ball in the center; in the expanded phase, the coloring matter is distributed throughout the now almost disk-shaped chromatophore.

The changes from one color phase to another are extremely rapid: in one common species of cuttlefish, the complete process from maximum contraction to maximum expansion was found to take exactly

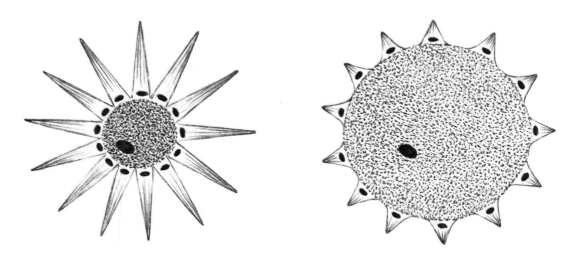

Chromatophore of the squid Loligo. These chromatophores actually alter their size and shape to effect the color change.

two-thirds of a second. Although this is a record even for cephalopods, the speed of the cell changes explains why the color shifts in these animals are so mobile that they pass over the body likes waves. One end of the cephalopod may still be pale while the other has already darkened or assumed an altogether different hue.

As indicated by the rapidity of these changes, they are all under direct nervous control. The animal's nervous system has three color centers, one of which seems to be a kind of "control center" capable of overruling the others. Blood-borne substances are also involved in some cases, but play a secondary role.

An angry or excited octopus may quickly assume a number of different shades ranging from orange and red to dark purple.

Frightened, an octopus may try to hide by changing its color to blend into the background.

The color changes in the chromatophores of most cephalopods are usually enhanced and emphasized by the light effects produced by the iridocytes. These cells form a firmly established base over which the dark brown or blackish melanophores and the lipophores, with their reddish yellow or reddish blue pigments, contract or expand to create any number of color combinations. The iridocytes produce not only white, but also pearly, iridescent interference colors in accordance with the principle of the colors of thin films. This explains why a cephalopod in an excited mood may present a veritable rainbow of rapidly shifting hues, a spectacle enhanced by the equally rapid changes in skin texture, which shifts from smooth to rough to warty and back again to smooth with a speed that matches the play of colors.

One of the most beautifully colored of all octopuses is the paper nautilus, or argonaut, found in the Mediterranean. This cephalopod is distinguished by the paper-thin outer shell of the female, as well as by the broadness of its first pair of arms. The male, on the other hand, looks very much like any other octopus and has no shell, a fact that was recognized only about a century ago, although the female paper nautilus has been known for two thousand years. Aristotle described this animal and handed down the fable on which its other name of argonaut is based. According to Aristotle, the paper nautilus uses its broad first pair of arms like sails, stretching them out of the water into the air so that it can be propelled along by the wind. In this fantastic position, the paper nautilus allegedly reminded the Greeks of Jason and his band of followers sailing to Colchis for the Golden Fleece in the ship Argo. The peculiar appearance and beautiful coloring of this octopus were undoubtedly partly responsible for that fanciful tale.

The upper parts of the paper nautilus are normally pistachio green, and its lower parts, brownish washed with silver. These background colors are speckled with tiny yellow and red dots, some of the larger ones bordered by a silvery rim, giving the animal a delicate tint of rosy silver. As the argonaut moves, the iridescent colors change, and if it gets excited, the pigment colors also change, resulting in hues of red, copper, and lilac that appear and disappear, rippling like waves over the body and shifting with the movements, as well as with the moods, of the animal.

From all observations of cephalopods, it seems clear that the color changes in these animals serve primarily as expressions of moods and emotions; their camouflage value seems to be of relatively minor importance.

Although much remains to be learned about the color changes in octopuses and other cephalopods, the research of the past decades has shed light on quite a few of the factors that contribute to the phenomena. It seems clear that the main stimuli triggering the changes are visual: the octopus gets excited by what it sees, and expresses this excitement by way of color change. The eyes of these animals are the most highly developed organs of sight found in any invertebrate, and they have excellent color vision. As soon as the eye sends the stimulus to the color centers in the brain, the message is passed on via the nervous system to the chromatophores at an incredible rate of speed.

Tests with one-eyed octopuses have shown that the color will change over the entire body area despite the animals' visual handicap, but that the changes on the "blind" side are much less striking. However, even complete blinding cannot inhibit color changes altogether, which proves that there are sensors in the skin that react to light and trigger a shift in hues. In addition, tactile stimulation also has been found to play a part.

Complex and fascinating, the diversity of both the mechanics of the color changes and the purposes for which this "color code" is employed by various marine creatures pose a challenge for those biologists intent upon unraveling the behavioral patterns of the animals that inhabit the vast underwater world.

Bibliography

Cochran, Doris. *Living Amphibians of the World*. Doubleday & Co., 1961.

Eibl-Eibesfeldt, Irenaeus. *Land of a Thousand Atolls*. University of Michigan Press, 1959.

Parker, George H. *Animal Color Changes and Their Neurohumors*. University of Cambridge Press, 1948.

Parker, George H. *Color Changes of Animals in Relation to Nervous Activity*. University of Pennsylvania Press, 1936.

Portmann, Adolf. *Animal Camouflage*. University of Michigan Press, 1959.

Pycraft, W. P. *Camouflage in Nature*. Hutchinson & Co., 1925.

Schmidt, Karl P. and Robert F. Inger. *Living Reptiles of the World*. Doubleday & Co., 1957.

Stephenson, E. H. and Charles Stewart. *Animal Camouflage*. Charles Black, 1946.

Index

Pages on which illustrations appear are shown in *italics*